INTEGRATED WATER DEVELOPMENT

Water Use and Conservation Practice
in Western Colorado

by

James L. Wescoat, Jr.
The University of Chicago

THE UNIVERSITY OF CHICAGO
DEPARTMENT OF GEOGRAPHY
RESEARCH PAPER NO. 210

1984

Copyright 1984 by James L. Wescoat, Jr.
Published 1984 by the Department of Geography
The University of Chicago, Chicago, Illinois

Library of Congress Cataloging in Publication Data

Wescoat, James L., Jr.
  Integrated water development.
  (Research paper; no. 210)
  Bibliography: p. 227
  1. Integrated water development—Colorado. 2. Water use—Colorado. 3. Water conservation—Colorado. I. Title. II. Series: Research paper (University of Chicago. Dept. of Geography); no. 210.
H31.C514 no. 210 910s [333.91'009788] 83-24130 [TC424.C6]
ISBN 0-89065-115-9 (pbk.)

Research Papers are available from:
The University of Chicago
Department of Geography
5828 S. University Avenue
Chicago, Illinois 60637
Price: $8.00; $6.00 series subscription

CONTENTS

LIST OF FIGURES v

LIST OF TABLES vii

ACKNOWLEDGMENTS xi

Chapter

    I. INTRODUCTION 1

   II. GEOGRAPHICAL CONTRIBUTIONS TO THE CONCEPT
       OF INTEGRATED WATER DEVELOPMENT 7

  III. A GEOGRAPHY OF WATER CONTROL ON THE COLORADO RIVER
       IN WESTERN COLORADO 25

   IV. PATTERNS OF WATER RIGHTS APPROPRIATION 59

    V. PATTERNS OF WATER DIVERSION 89

   VI. CHANGING PATTERNS OF WATER USE 109

  VII. WATER CONSERVATION PRACTICE IN WESTERN COLORADO . . . 127

 VIII. THE ORGANIZATIONAL FABRIC OF WATER MANAGEMENT 161

   IX. CONCLUSION 183

APPENDIX 193

BIBLIOGRAPHY 227

LIST OF FIGURES

1. Map of Colorado River Basin and the Study Area 2
2. Diagram of Geographical Contributions to Water Resources Management 20
3. Map of Water Development in the Colorado River Basin . . . 27
4. Map of Upper Basin Water Development and Oil Shale Resources 37
5. Map of Water Administration in Colorado 41
6. Map of Water Division 5 and Transbasin Diversions 43
7. Map of the Study Area and Substate Water Districts 47
8. Map of Stream Drainages in the Study Area 61
9. Timing of Absolute Flow Decrees 66
10. Intensity of Absolute Flow Rights by Stream Basin 73
11. Intensity of Conditional Flow Rights by Stream Basin . . . 76
12. Absolute Storage Rights by Stream Basin 79
13. Conditional Storage Rights by Stream Basin 80
14. Map of Oil Shale Project Lands in the Study Area 83
15. Intensity of Administrative Record-Keeping 93
16. Intensity of Recorded Diversions 99
17. Map of Irrigation Application Rates 102
18. Map of Mean Basin Ranks 104
19. Map of Water Rights Changes by Stream Basin, 1970-1977 . . 117
20. Diagram of Relative User Locations in a Stream Basin . . . 131
21. Irrigation Application Rates in Western Colorado, 1945-1979 133

22. Map of Major Canals and Sample Survey Sites
    in the Grand Valley Area 138
23. Annual ACP Expenditures in Mesa County, 1977-1980 150
24. Map of ACP Projects in Grand Valley, 1980 151
25. Annual ACP Expenditures in Garfield County, 1980 155
26. Map of ACP Projects in Garfield County, 1980 156
27. Comparison of Water Organizations in Colorado
    and the Colorado Portion of the Upper Colorado
    River Basin, 1969 164
28. Irrigated Acreage by Type of Organization
    in Colorado, 1910-1978 170
29. Incorporation of Water Organizations in Garfield County . 171
30. Incorporation of Water Organizations in Mesa County . . . 171

LIST OF TABLES

1. Stream Basins Included in the Analysis 60
2. Correlation Coefficients for Decree Types 69
3. Absolute Flow Rights by District (AFLOW) 72
4. Intensity of Absolute Flow Rights (AFLOW/Mi^2) 72
5. Conditional Flow Rights by District (CFLOW) 75
6. Intensity of Conditional Flow Rights (CFLOW/Mi^2) 75
7. Absolute Storage Rights by District (ASTOR) 77
8. Intensity of Absolute Storage Rights (ASTOR/Mi^2) 78
9. Conditional Storage Rights by District (CSTOR) 78
10. Intensity of Conditional Storage Rights (CSTOR/Mi^2) . . . 81
11. Prices for Reudi Reservoir Water 85
12. Correlation between Water Rights and Diversion Variables . 97
13. Comparison of Total Diversions and Total Absolute Flow Rights 97
14. Summary of Water Rights Changes by Districts 114
15. Water Supply Organizations and Major Canals in the Grand Valley Area 137
16. Projected Levels of Land Treatment on Irrigated Land Through a Ten-Year Continuation of Regular Program Activity 143
17. Estimated Cost of Conservation Improvements 144
18. Percentage of ACP Payments in the United States by Practice Category, 1940-1979 149
19. Maximum Federal Cost-sharing Assistance by Water Conservation Practice, Grand Valley Salinity Control Program, 1981 153

20.	Benefits and Costs of ACP Water Conservation Assistance, 1975-1979	157
21.	Special District Classification in Colorado	167
22.	Summary of Decreed Water Rights	195
23.	Mean Values of Basin Water Rights Variables by District	196
24.	Correlation Coefficients for Absolute Flow Rights by District (AFLOW)	196
25.	Correlation Coefficients for Area-Adjusted Absolute Flow Rights (AFLOW/Mi^2)	197
26.	Correlation Coefficients for Conditional Flow Rights by District (CFLOW)	197
27.	Correlation Coefficients for Area-Adjusted Conditional Flow Rights (CFLOW/Mi^2)	198
28.	Correlation Coefficients for Absolute Storage Rights by District (ASTOR)	198
29.	Correlation Coefficients for Area-Adjusted Absolute Storage Rights (ASTOR/Mi^2)	199
30.	Summary Statistics for District Water Diversions, 1980	200
31.	Correlation Coefficients for Percentage of Structures Recorded (%REC)	201
32.	Correlation Coefficients for Total Diversions (CFS)	202
33.	Correlation Coefficients for Area-Adjusted Diversions (CFS/Mi^2)	203
34.	Correlation Coefficients for Irrigated Acreage (AC)	204
35.	Correlation Coefficients for Length of Irrigation Season (DAYS)	205
36.	Correlation Coefficients for Irrigation Application Rates (AF/AC)	206
37.	Correlation Coefficients for Seniority (RANK), 1980	207
38.	Correlation Coefficients for All Water Rights Changes (ALLCHG)	208
39.	Correlation Coefficients for Contested Water Rights Changes by District (CONCHG)	209
40.	Mean Values for Annual Diversion Records and Independent Variables	210
41.	Trends in Water Diversion and Independent Variables, 1960-1978	211
42.	Correlation of Water Use Variables with Price, Income, and Precipitation Variables	212
43.	Correlation Results for Annual Diversion Records (Lag 1)	213

44. Land Use Summary for the Grand Valley Area by Canal Service Area, 1969 214

45. Joint Frequency Distribution of Field Size by Canal Service District 215

46. Frequency Distribution of Existing Flow Measurement Systems by Canal Service District 216

47. Frequency Distribution of Existing Pipeline Systems by Canal Service District 217

48. Frequency Distribution of Existing Concrete Ditch Lining by Canal Service District 218

49. Frequency Distribution of Expressed Flow Measurement Needs by Canal Service District 219

50. Frequency Distribution of Expressed Pipeline Needs by Canal Service District 220

51. Frequency Distribution of Expressed Ditch Lining Needs by Canal Service District 221

52. Frequency Distribution of Expressed Land Leveling Needs by Canal Service District 222

53. Frequency Distribution of Expected Future Land Use by Canal Service District 223

54. Salinity Control Cost-sharing Assistance by Canal Service District, 1980 224

55. Frequency Distribution of Actual Conservation Expenditures by Canal Service District, 1980 225

ACKNOWLEDGMENTS

This research on changing patterns of water use and conservation practice in western Colorado has been carried out under generous support from the National Science Foundation and the National Institute for Socioeconomic Research though the views and policies of those organization are not necessarily represented herein. The Institute, in particular, provided an educational grant for field work and data collection in the study area during the summer of 1981. I owe an initial debt to Professor R. Stephen Berry for introducing me to the research problem. Special acknowledgment is owed to geography professors Norton S. Ginburg, Chauncy D. Harris, and James T. Meyer of the University of Chicago for giving guidance throughout the project. Numerous public officials and water users were generous with both their valuable time and information. Elizabeth Brooks provided creative assistance in preparation of the figures. Finally, research scheduling and progress were firmly yet beneficently supervised by my wife Florrie and daughter Ruth.

CHAPTER I

INTRODUCTION

Increasing competition for water in the Upper Colorado River Basin has generated a wave of concern regarding the adequacy of existing water supplies, the need for new reclamation projects, and the future of traditional water use practice and lifeways in the region. In western Colorado concern has revolved around water demand projections for oil shale, coal, and associated urban development. Although by no account the first set of conflicts over waters of the Colorado River, often heralded as "the most developed and litigated river" in the world, these issues do represent an important and unfinished chapter in the development of the Upper Basin; for up until the middle of this century the Western Slope of Colorado was relatively free from the types of internal pressures that had driven southern California, central Arizona, and the Front Range of Colorado toward grandiose schemes of water acquisition (figure 1).[1]

Divergent water demand forecasts and conflicting opinions on water management choices have fed confusion about the problems which face western Colorado. Covetous threats from Lower Basin and Front Range water seekers, for example, have fueled a perceived need in western Colorado to develop all compact-apportioned waters before considering a broader range of management approaches.[2] Conservation can come later--the argument goes--after compact waters are securely developed. The notion of "maximum utilization" sounded in recent Colorado Supreme Court decisions, however, provides a more balanced

---

[1]. The popular expressions "Western Slope" and "Front Range" refer to the hydrologic and historically adversarial bifurcation of Colorado by the Continental Divide. The Western Slope drains western Colorado into the Colorado River and its tributaries, on the one hand, and the Front Range drains eastern Colorado into the Platte, Arkansas, and Rio Grande river basins, on the other.

[2]. The expression "compact-apportioned share" refers to water allocated under the Colorado River Compact of 1922 (42 Stat. 171, 46 Stat. 3000) and the Upper Colorado River Compact of 1948 (63 Stat. 31).

Fig. 1. Map of the Colorado River Basin and the study area

view of vested rights, conservation needs, and the public good.3 It is not at all certain that single-minded development of Colorado's compact-apportioned waters is either an efficient or desirable path toward "maximum utilization."

Haphazard interpretations of recent shifts in land and water use patterns also cloud the problems faced in water management. Arguments that industrial water demand is "drying up" agricultural lands meet counter-arguments that changes in water use represent only the indirect effects of shifts in the demand for agricultural land and the declining viability of agricultural production in western Colorado.

Recent regional water demand forecasts identify no important impacts of energy development on existing patterns of water use (assuming development of compact-apportioned waters and free transfer among competing water users). These conclusions have led some to question the existence of any real water policy issues in western Colorado. These forecasts also suggest that water conservation is not a feasible path toward meeting new water demands. Opponents argue that conflicts may still be expected in local areas and across jurisdictional boundaries due to constraints on water distribution and transfer; they point to excessive irrigation rates, water waste, and river salinity. Because the scale of analysis in these studies is generally fixed on regional basin-wide problems, differentiation among varieties of local water use trends, opportunities, and problems tends not to occur. The confusion then spills over into policy recommendations--water zoning, growth management, reservoir development, water transfer rules, and tax measures--often designed for local implementation. Occasionally, the debate leads to wider critiques of Colorado water law, calls for repeal of the appropriation doctrine, and ambitious but vaguely sounded pleas for water policy reform.

Study Approach

Starting out with the recent optimistic forecasts and laissez-faire conclusions of regional water investigations, the goal of this research is to discover whether a more detailed geographical analysis of local water use can lead to a broader range of choice among paths toward "maximum utilization." The following questions

---

3. *Fellhauer v. People,* 167 Colo. 320, 447 P.2d 986 (1968) introduced the concept of "maximum utilization." Although *Fellhauer v. People* involves a dispute over regulation of groundwater withdrawals and not compact issues per se, the concept of maximum utilization has broader application in water resource management. See, for example, *A-B Cattle Co. v. United States,* 196 Colo. 539, 589 P.2d 57 (1978) in which the Court built upon the *Fellhauer* case in refusing to protect inefficient water use practices.

are addressed: what is the areal configuration of water supply problems (e.g., over-appropriation, scarcity, waste, and conflict); how are these problems related to opportunities and constraints on water conservation; and finally, under what circumstances might improvements in irrigation efficiency be able to occur in conjunction with industrial water development?

The argument that a geographical approach can contribute in some fashion to the identification of water policy issues, to the evaluation of water management decisions, and to the discovery of linkages between complementary or successive water uses is critically assessed in chapter two. It is frequently stated, for example, that the descriptive synthesis of early river basin studies and the broadened decision-making framework of natural hazards research represent an "integrated" approach to water resources management.

In chapter two the concept of "integrated water development" is taken up directly. Integrated development is defined as a process of adjustment in water use patterns and practices that results from the search for cooperative linkages among water users. Evaluations of "free-market" transfer among competitors, by contrast, have focused almost solely on the relative efficiency of water transfers from one use to another, often overlooking strategies for making concurrent improvements in water use efficiency. For purposes of illustration, several examples of integrated water development are briefly presented, followed by a problem statement for the geographical assessment of industrial and agricultural water development in western Colorado.

The "nested hierarchy" of water control institutions that affect water use in western Colorado is outlined in chapter three--proceeding from international agreements with Mexico, to Federal and interstate policies, State water administration, and ultimately to sub-state water control activities. The objective is to sort out institutions and policies in larger arenas that influence local water practice, as well as those that encourage or impede the process of integrated water development.

After identifying water management controls on the Upper Colorado River main stem, attention is directed toward the principal research problem--local water practice in Garfield and Mesa counties in Colorado. In chapter four the cumulative record of water rights in these two counties is assessed, with special attention given to the differences between patterns of water appropriation on the main stem and on the tributaries of the Colorado River. Cumulative patterns of water rights appropriation are analyzed at the stream basin and water district levels, taking into account local differences in physiography and relative location.

In chapter five the legal record of water appropriation is then contrasted with records of actual water diversions in order to evaluate claims of over-appropriation, inefficient irrigation practices, and the profusion of "paper rights" in western Colorado. The incomplete coverage of diversion records also permits inferences to be made about deficiencies in the public administration of water rights and water conflicts.

Recent changes in both water rights and water diversions are evaluated in chapter six. Water court records provided detailed information on the character and distribution of conflict accompanying water rights changes. Trends in water use are identified, and local irrigators' responses to annual variation in precipitation, farm income, and product prices are assessed. Interviews with water managers, state officials, water professionals, and energy firms aided in the interpretation of data sources and trends.

Regional studies of the Upper Colorado River Basin have downplayed the potential contribution of conservation to meeting water demand in energy growth areas. In chapter seven the limiting assumptions behind these previous studies are pointed out, and the case is reopened for integrating conservation needs with water development objectives. The initial task lay in documenting current conservation practice and perceived conservation needs in the study area. Conservation is not a simple phenomenon, but rather one that is frequently motivated by non-conservation objectives, such as lower costs, improved farm operations, or improved yields. The second research task in chapter seven involved a comparison of expressed conservation needs with actual patterns of conservation investment. Investment patterns provided insight into the potential adoption of conservation practices, the types of water uses that tend to be improved, the user groups involved, and their locational circumstances.

Historical assessments of reclamation practices have increasingly laid emphasis on the role of local organizations in stimulating the development and subsequent operation of major water systems. This constitutes a reaction to a longstanding focus on the Federal role in water resources development. Equally important, however, are the relationships among different levels of government and among different types of water organizations for either facilitating or constraining various water management activities. For water resources, more so than land, the organizational fabric of management strongly influences the prospects for conservation and cooperative resource development.

In chapter eight, therefore, the evolution of water supply organizations in western Colorado is reviewed. In many cases the organizations presently operating were initially established for purposes other than their present functions; in other cases their articles of incorporation limit the type and location of activities that may be pursued. The principal research questions in this chapter are: first, in what ways have local water organizations adapted to changing needs; and second, in what ways do the geographical circumstances of existing organizations bear on the feasibility or, indeed, desirability of the integrated development process presently under consideration? Profiles of organizational adjustments which facilitate conservation are then constructed for sub-basins in the study area, based on the findings in chapters four through seven.

The conclusions of this research project are several in nature. First, the explicitly geographical approach used to describe water use and conservation practice is critically assessed. Second, key administrative and water supply problems in the study are summarized. Finally, the concept of integrated water development is evaluated for its applicability to western Colorado, with particular emphasis given to institutional arrangements facilitating water conservation and exchange. This inquiry consists of a sequence of related tasks rather than a single task directed at a single problem. Each chapter raises questions for further development while serving as a foundation for subsequent chapters. Conclusions, therefore, are intended to have programmatic value of a general nature as well as practical application to the transition in water management currently under way in western Colorado.

CHAPTER II

GEOGRAPHICAL CONTRIBUTIONS TO THE CONCEPT OF INTEGRATED WATER DEVELOPMENT

In his recent book, *A River No More: The Colorado River and the West,* Phillip Fradkin starts out with the question, "Why another book on the Colorado River?"[1] Those who have followed the false starts in oil shale development over the past sixty years, the perennial media coverage of water supply issues in western Colorado, and the profusion of technical studies on the potential impacts of energy development on water availability might ask a similar question, "Why another study of water and energy in western Colorado?" Two responses are offered to this question: first, there are substantive omissions and deficiencies in past research on water management in western Colorado; and second, geographical contributions to water resource research offer a creative framework for resolving these deficiencies. The Introduction cited as deficiencies the inattention of past studies to local variation in water management problems; the neglect of historical patterns of water use, water rights changes, and administration; and the dismissal of improved water use efficiency as a feasible path toward "maximum utilization." In the Introduction it was suggested that geographical concepts and methods offer an integrated framework for resource analysis. This chapter assesses the accomplishments and unfulfilled agenda of integrated water development, as defined by geographers and others, in order to establish a fresh conceptual and methodological foundation for proceeding on yet another study of energy and water in western Colorado.

The term "integrated development" emerged in close association with the advent of river basin planning in the 1930's but was not precisely defined in specific legislation or policies. Rather, it generally designates the simultaneous consideration of formerly

---

1. Phillip L. Fradkin, *A River No More: The Colorado River and the West* (New York: Alfred A. Knopf, 1981), p. xvii.

separate resource development problems. Thus, it serves as a programmatic expression for broadening the scope of planning investigations and is not an established planning technique.[2] Geographical contributions to the concept of integrated development began in the early period of river basin development planning and were recast in broader terms in Gilbert White's six-fold partitioning of water resources geography.[3] Reappraisal of these accomplishments in light of the present research task suggests several important extensions and modifications of the original concept as well as more explict formulation of what the geographer can say about it.

Origins in River Basin Development

River basin development and the science of hydrology have an extended history, but the concept of a drainage basin as a planning region was not formally developed until this century.[4] River basin planning represented a break with traditional engineering practice which had dealt for the most part with single objectives (e.g., flood control, power, or irrigation) and a narrow range of technological alternatives for meeting those objectives. River basin planning sought to encompass multiple objectives for water use and a full range of technological and policy actions for implementation. Thus, it was said to represent an integrated approach to natural resource development; but what, specifically, was meant by the expression integrated? An early paper on river basin development by White lays out the major themes:

1. A shift from single-purpose to multiple-purpose storage projects, i.e., *integration of facilities functions*
2. A shift from river channel engineering to a basin-wide framework for water management, i.e., *integration of related hydrologic processes within the scope of planning*--initially

---

2. Cf. "comprehensive plans," "coordinated joint plans," "multiple use managment," and other institutionally defined planning practices. Beatrice Holmes, *A History of Federal Water Resources Programs, 1800-1960*, U.S. Department of Agriculture, Miscellaneous Publication, no. 1233 (Washington, D.C.: Government Printing Office, 1972) associates the term integrated development with New Deal water planning. In later years, the idea of integrated development as a planning concept would be formalized as "comprehensive river basin planning" in the Water Resources Planning Act of 1965 (79 Stat. 245, 42 U.S.C. 1962). See Beatrice Holmes, *History of Federal Water Resources Programs and Policies, 1961-1970*, U.S. Department of Agriculture, Miscellaneous Publication, no. 1379 (Washington, D.C.: Government Printing Office, 1979), chap. 10. The expression integrated river basin development has perhaps wider currency among international practitioners, e.g., United Nations, Department of Economic and Social Affairs, *Integrated River Basin Development* (New York, 1958).

3. Gilbert F. White, "Contributions of Geographical Analysis to River Basin Development," in *Readings in Resource Management and Conservation*, ed. Ian Burton and R.W. Kates (Chicago: University of Chicago Press, 1965), pp. 375-394.

4. Ludwig A. Teclaff, *The River Basin in History and Law* (The Hague: Martinus Nijhoff, 1967).

precipitation and runoff processes; and subsequently groundwater, soil moisture, and weather

3. A shift toward large-scale public investment in water projects as a means of regional economic development, i.e., *integration of economic and engineering objectives.*

White noted that several other ideas had been associated with river basin development, but that due to a number of important limitations, they had generated as much controversy as currency:

1. *Integration of water resources and land use planning*
2. Experiments in unified basin administration such as the TVA, the Compagnie National du Rhone, and the Damodar Valley Authority, i.e., *integration of administrative organizations and functions*
3. Negotiation of international and interstate agreements for the management of shared resources, i.e., *political and territorial integration through river basin development.*[5]

In a review over fifteen years later White would include among these an important addition:

4. The trend toward public involvement in decision-making and the consideration of perception, attitudes, and opinions in the process of decision-making, i.e., *integration of political participation and behavioral considerations in the planning process.*[6]

For geographers these advances initally came together in systematic basin surveys and development plans. The early surveys provided a succinct, if sometimes static, appraisal of opportunities and constraints on resource development. Investigation of specific development problems, such as rural settlement, flood risks, infrastructure needs, land use change, and water distribution, was generally carried out within the methodological framework of regional description and synthesis that was characteristic of mainstream geographical studies.[7] A variety of institutional entities ranging widely in jurisdiction and powers emerged to

---

5. Gilbert F. White, "A Perspective of River Basin Development," *Law and Contemporary Problems* 22 (1957): 157-184.

6. Idem, "Role of Geography in Water Resources Management," in *Man and Water*, ed. L. Douglas James (Lexington, KY: University of Kentucky Press, 1974), pp. 102-21.

7. The earliest Federal basin studies, "308 reports," followed from the Rivers and Harbors Act of 1925 (43 Stat. 1186) and House Document No. 308 (69th Congress, 1st Session, 1926); the 1927 Rivers and Harbors Act (44 Stat. 1015) authorized the Corps of Engineers to conduct the studies. The summary documents from these studies reflect the central role of geographers in basin surveys and policy analysis--see President's Water Policy Commission, series title, *Ten Rivers in America's Future* (Washington, D.C.: Government Printing Office, 1950). A geographical approach was also employed by the Bureau of Reclamation--exemplified in the *Columbia Joint Investigations,* 24 vols. (Washington, D.C.: Government Printing Office, 1941), directed by Harlan H. Barrows.

produce, maintain, and coordinate basin plans. Subsequent national water assessments in 1968 and 1978 by the Water Resources Council followed a similar scheme, dividing the nation into 17 large hydrologic regions for data collection and problem analysis.[8]

Initially it was thought that the river basin could serve as an areal framework for comprehensive resource management and economic development as well as for water resources planning. The conceptual pitfalls of conflating regional problems, however, were clearly spelled out by geographers as early as 1935:

> The worst possible kind of region [for regional planning] would be a drainage basin, which, as W.M. Davis pointed out long ago, is seldom in any sense (except drainage) a unit. (Preston E. James)
>
> The river basin has the advantage of being easily and sharply bounded, but it has the much greater disadvantage of using a criterion which does not provide homogeneity in major problems. The lower courses of two river systems are likely to have more in common than the upper and lower reaches of the same system. (Robert S. Platt)
>
> It seems to me almost obviously absurd to take a river basin as a fundamental regional basis. Take, for example, the Mississippi River or the Ohio Basin; how can one find any regional similarity on such a basis? (Richard Hartshorne)
>
> . . . except for engineering works, the drainage basin is the poorest type of region to deal with . . . I am not in favor of using the watershed for any other type of planning than engineering plans dealing with power, navigation, and flood control, because this geographic peculiarity has not been as determinant of homogeneous areas as have soil, climate, and type of original settlement. (T.J. Woofter)[9]

Integration of water resources and land use planning was a projected but unrealized goal of river basin planning. This has been due in large part to non-conterminality among resource distributions and their associated management problems. It is further complicated by the different systems of property that apply to the use of land and water resources. John Borchert expressed this well for the Wabash Basin:

> Moreover, different problems usually have different geographical boundaries. For example, the Wabash Basin may be troubled by flooding, erosion, lack of industrialization, a heavy outmigration, declining farm production, and blighted towns. While all these things come together in one area and might be considered to comprise the Basin's "problems," the meaningful geographical areas coincident with each of these problems encompass different territories--some cover only a portion of the Basin while others include territory well beyond the basin boundaries.[10]

---

8. Federal involvement in Title II River Basin Planning and national assessments, authorized by the 1965 Water Resources Planning Act (79 Stat. 244, 42 U.S.C. 1962), is projected to be dramatically reduced and the Water Resources Council dissolved in 1982 making this an appropriate time to assess the emergence and florescence of a strongly geographical mode of analysis.

9. U.S. National Resources Committee, *Regional Factors in National Planning and Development* (Washington, D.C, 1935), pp. 146-49. See also John Friedmann and Clyde Weaver, *Territory and Function* (Berkeley: University of California Press, 1979).

10. John R. Borchert, "Formulation of Regional Development Policy in the Midwestern Setting," in *Regional Development and the Wabash Basin,* ed. Ronald R. Boyce (Urbana, IL: University of Illinois Press, 1964), p. 108. On property rights considerations see J.H. Dales, "Land, Water, and Ownership," in *Economics of the Environment,* ed. Robert and N.S. Dorfman (New York: W.W. Norton and Co., 1977), pp. 229-44.

Efforts to integrate river basin development and economic development have had limited success for similar reasons. John Friedmann demonstrated in the Tennessee River Basin that river development contributed not to basin-wide growth but rather to the acceleration of urban and industrial trends already underway.[11] Reclamation programs aimed at regional economic development have received criticism for failing to consider interregional impacts, for misconstruing forecasts of benefits and costs, and for deterministically representing the relationship between water development and economic growth. Economists have argued forcefully, for example, against Federal irrigation policies in the western United States, attempting to demythologize what one has termed the "water-is-different" syndrome, and to substitute in its place a more classical view of water resources as ordinary production inputs and consumer goods.[12] Two major innovations in water management thus came to dominate discussion concerning integrated river basin development: first, application of an efficiency criterion to basin-wide development projects, and second, consideration of a wider range of alternatives for surface water management.

Given the early recognition of theoretical weaknesses in certain integrated development concepts, their persistence in water resources and planning efforts becomes problematic. To answer the question of how such ideas have persisted, evidence may be sought in 1) the pattern of unfulfilled planning agendas, 2) the political grounds for integrated basin development programs, and 3) what might be called the fallacy of the "water-is-no-different syndrome."

The obvious successes of river basin planning lie principally in improved engineering practice, e.g., multiple purpose facilities planning and basin-wide hydrologic modeling. International and interstate basin agreements offer a more varied record of success and failure, demonstrating cooperative behavior among general purpose governments if not always achieving reconciliation in cases of extreme conflict.[13] The most consistent criticism seems to be

---

11. John R. Friedmann, *The Spatial Structure of Economic Development in the Tennessee Valley,* Research Papers, no. 39 (Chicago: University of Chicago, Department of Geography, 1955).

12. The "water-is-different" phrase was coined by Maurice M. Kelso, "Competition for Water in an Expanding Economy," *Water Resources and the Economic Development of the West* 16 (1967): 187-196. Major critical pieces include Jack Hirshleifer, et al., *Water Supply: Economics, Technology, and Policy* (Chicago: University of Chicago Press, 1960); Earl O. Heady, et al., *Agricultural and Water Policies and the Environment* (Ames, IA: Center for Agricultural and Rural Development, 1972); Robert A. Young, "Economic Analysis of Federal Irrigation Policy: A Reappraisal," *Western Journal of Agricultural Economics* 3 (1978): 257-267; and Maurice M. Kelso, W.E. Martin, and L.E. Mack, *Water Supplies and Economic Growth in an Arid Environment* (Tucson: University of Arizona Press, 1973).

directed toward river basin planning, per se,
1. for failing to produce implementable plans
2. for failing to guide the policy-making process
3. for subsuming problems in land use planning and economic development that are affected by, but not properly managed as part of, the surface water system.

In defence, it should be pointed out that authorization of engineering projects has often preceded comprehensive basin surveys. This results in an unfortunate development sequence in which technological alternatives are selected first, after which planning becomes an exercise in adjusting to the new physical system.[14] A recent study for the Water Resources Council argued that Federal Level B river basin planning should not purport to produce implementation plans, but should rather capitalize on its demonstrated benefits, " . . . improved communication among agencies; citizen involvement in the consideration of resource issues; increased dissemination of information; and State involvement in Federal decision-making for water resources."[15]

This last point on Federal decision-making introduces a more sweeping query on the political grounds of integrated water development in the United States. The integrated development concept evolved in response to perceived national water problems and was promoted by the Federal government--a weighty participant in interstate compacts as well as in most major water development schemes.[16]

Occasionally this behavior was emulated in river basin planning at the State level. The concept of integrated water development tends to weaken in local applications, however, in large part due to the property rights conflicts mentioned above. Not only do land and water rights have important differences, but components of the hydrologic system, such as streamflow, precipitation, soil

---

13. J.D. Chapman, *The International River Basin* (Vancouver: British Columbia Press, 1963); W.R. Derrick Sewell and Gilbert F. White, "The Lower Mekong, an Experiment in International River Development," *International Conciliation* (1966): 5-63; W.R. Derrick Sewell, "The Mekong Scheme: Guideline for Solution to Strife in Southeast Asia," *Asian Survey* (1968): 448-95. The lengthy negotiations leading to apportionment of Colorado River waters is given detailed treatment in Norris Hundley, Jr., *Dividing the Waters* (Berkeley: University of California Press, 1966).

14. On the Columbia River, for example, construction of Grand Coulee Dam began in 1934, the Columbia Basin Project Joint Investigations were completed in 1941, yet a comprehensive basin survey was not completed until 1947.

15. Ralph M. Field Associates, Inc., *Regional Water Resources Planning*: *A Review of Level B Study Impacts for the U.S. Water Resources Council* (Washington, D.C., 1979).

16. Bethemont similarly notes the dominant role of national governments in the water management activities of socialist countries (such as Rumania and Hungary), in Jaques Bethemont, *De l'eau et des hommes: essai géographique sur l'utilization des eaux continentales* (Paris: Bordas, 1977), p. 127.

water, non-tributary groundwater, and storage water, are in many States administered under separate systems of public and private ownership. In general, property rights to land are fixed in place, easily measured, and publicly recorded, whereas surface water rights (in the appropriation states) are usufructary, successive in time and space, and not as precisely documented; groundwater and riparian rights are still less well-defined; and precipitation rights will probably be fought over in future decades. Moreover, the fact that land use controls generally arise from State enabling legislation to local governments, whereas water rights are administered by the State directly, further impedes integrated land and water management at either level of government.

The point to be underscored here is that the concept of integrated water development was conceived with high ambitions but relatively shallow, if broad, political grounding in the Federal government. The significance of local water insititutions and practice has been consistently underplayed, giving evidence in American water planning to that half of the Wittfogel fallacy asserting the importance of bureaucratic control over large water management systems.[17] A recent comparative study of irrigated areas in the western United States goes a long way toward remedying this view by observing that local water organizations have historically exercised great influence over the scope of water projects, their operation and maintenance, and the terms for their repayment.[18] Former Water Resources Council chairman Henry Caulfield has called for a critical reappraisal of the Federal role in water management citing the efficacy of State and local institutions for handling most water problems--a claim that may be tested in the coming years.[19]

Public participation programs do not speak to the central issues of this critique. Public participation has taken several forms in basin development programs--such as the required demonstration of "local support" for Army Corps of Engineers projects; "grass roots" involvement in the TVA; and more formal

---

17. Karl Wittfogel, "The Hydraulic Civilizations," in *Man's Role in Changing the Face of the Earth,* ed. William L. Thomas, Jr. (Chicago: University of Chicago Press, 1965), pp. 152-65.

18. Arthur Maass and Raymond L. Anderson, *. . . and the Desert Shall Rejoice: Conflict, Growth and Justice in Arid Environments* (Cambridge, MA: MIT Press, 1978).

19. Henry Caulfield, "Let's Dismantle (Largely but not Fully) the Federal Water Resource Development Establishment: The Apostasy of a Longstanding Water Development Federalist," in *Water Needs for the Future,* ed., Ved P. Nanda (Boulder, CO: Westview Press, 1977), pp. 171-78; and see Frank J. Trelease, "The Model Water Code, the Wise Administrator and the Goddam Bureaucrat," *Natural Resources Journal* 14 (1974): 207.

citizens' workshops, public participation programs, and other vehicles for public review of proposed actions--all of which aim at vertical political integration as a means of obtaining consensus or conflict resolution. A United Nations report in 1958 reflects the underlying logic and separation of interests in vertically integrated resource development:

> An integrated river basin development programme will usually include large hydroelectric plants with interconnecting transmission lines, and major flood protection and irrigation works, which will be built and operated by the central authority. It will also include numerous watershed management improvement works, farm irrigation and drainage systems, and small-scale industrial enterprises which will not only have value in themselves, but will contribute to the success of the over-all programme. It is here that citizen participaton can be of the greatest importance.[20]

Considerable discussion has been devoted to the relationship between central and local governments in water control. Of principal importance here is the observation that integrated water development is historically rooted in Federal water policies and not in the cooperative behavior of local water users. It is not surprising then that the major accomplishments of these water policies occur for the most part in larger political arenas and in improved engineering practice.

While it may be useful to think of water as any other good for economic analysis and to strip away the "water-is-different" arguments for public spending decisions, much has been lost and smoothed over in the demythologizing of water. Water truly is different from other resources by virtue of its fluid behavior, geoecological linkages, perceived cultural value, and at the same time its susceptibility to human management. In a geographical study of irrigation in Utah Valley, James Hudson observed that,

> Most farmers in the valley do not view water as simply another input, such as fertilizer . . . In many cases, the water stock is part of an inherited estate, and it becomes an "heirloom" . . . Water's role in agriculture is easily overvalued: statements are often heard from farmers that "the land is worthless without [irrigation] water," even though some farms in the valley produce crops by dry farming or by fall sowing, and even though some land can be sold at high prices for suburban development. As a result, water is presently not regarded as a simple economic good, but as something removed from market transactions. Attempts to convert it to an economic good by selling surplus water to the highest desperate bidder are regarded locally as "water profiteering," while attempts to acquire large additional supplies are regarded as "water hogging."[21]

Moreover, as White put it, "No two rivers are the same."[22] Far from being an argument for geographical relativism, the key issue is rather that important geographical issues have fallen into neglect

---

20. United Nations, Department of Economic and Social Affairs, *Integrated River Basin Development* (New York, 1958).

21. James Hudson, *Irrigation Water Use in Utah Valley, Utah*, Research Papers, no. 79 (Chicago: University of Chicago, Department of Geography, 1962), pp. 229-30.

22. White, "A Perspective," p. 160.

which might strengthen the concept of integrated water development and clarify the research tasks of water resources geography.

The decline of geographical involvement in river basin development has been interpreted as a failure to directly address decision problems (a criticism also directed toward river basin planning generally):

> An explanation of why such geographical appraisals [land and water resource surveys] have not been emulated elsewhere must include personal administrative prejudice, but also must embrace the argument that unless they apply criteria and classes directly relevant to technical and economic decisions, they are fruitless for river planning purposes.23
>
> It would be misleading, however, to think of major programs of study looking to river basin design as ever being largely geographic . . . Commitment to such work does not necessarily involve geographers, for much of it is in borderline fields: hydrologists are involved in water balance study, soil scientists and foresters in land classification, economists in estimating external economies and so forth. Although geographers have done useful work which suggests they have a certain competence and may expand their services in those directions, there is no assurance that they will handle the work in the future.24

Given similar criticisms applied to river basin planning generally, one wonders if the value of geographical surveys has been accurately represented and, more importantly, why the warnings against using drainage basins for regional planning were not translated into alternative or more refined areal frameworks for regional water management. Instead of addressing these questions, geographers redirected their work to service problems in economic and technical decision-making, leading some to question what is "largely geographic," if anything.

Origins in the Analysis of Decision-Making Behavior

The commitment to structuring geographical research in terms of select decision parameters has endured, with few alterations in the basic approach, for thirty years. White classified geographical contributions to river basin planning and water resource management in the following manner:

1. Estimation of available resources
2. Expansion of the "range of choice" (among alternative actions, ectives, or plans perceived by decision-makers)
3. Assessment of technologies
4. Appraisal of economic efficiency
5. Assessment of spatial linkages
6. Assessment of social guides to decision.25

---

23. White, "Contributions of Geographical Analysis," p. 386.

24. Ibid., p. 393.

25. White, "Contributions of Geographical Analysis," pp. 375-394; White, "Role of Geography," pp. 102-121; and W.R. Derrick Sewell, "Geographical Research in Water Management in Canada," in *Water Management Research: Social Sciences Priorities,* ed. W.R. Derrick Sewell, et al. (Ottawa: Department of Energy, Mines and Resources, 1969). A similar framework is put forward for water planning generally in Dennis J. Parker and Edmund C. Penning-Rowsell, *Water Planning in*

The evolution of White's framework deserves special attention. It emerged concurrently with systems analysis and applied welfare economics in water resources research. Great hopes were pinned on the decision sciences for conflict resolution and project evaluation. Water resources geography adopted a systems approach, for example, in appraisals of project efficiency and technological alternatives, but it contributed primarily to data inputs and case studies as opposed to the development of theory or method. In 1953 White stated:

> Nothing less than a systems analysis of the whole complex of natural and social processes at work will yield in the long run a sound basis for decision, and it is toward this that we should move.[26]

A decade later, however, there were some questions. Systems analyst Myron Fiering stated, "Optimization seems almost to have become a new religion, with the applied mathematician its high priest and the automatic computer its holy scripture."[27] Finally, however, White reflected that, "It now seems likely that change in public preferences with respect to water quality or recreation or dependence on large dams has had far greater effect than has the refinement of economic analysis," and he suggested a critical appraisal of the research agenda.[28]

In response White and other geographers have focused on the behavioral basis for decision-making and the somewhat ambiguously labelled "social guides." The conceptual framework for this research finds its origins in models of perception and choice in the behavioral sciences.[29] Particular attention is given to specifying the full "range of choice" in water management and to interpreting the attitudes expressed toward those choices by various decision-making groups, including engineers, public officials, and user groups. Research results frequently identified misperceptions or biases that had unjustifiably limited the range of choice in

---

Britain, The Resource Management Series, 1 (London: George Allen & Unwin, 1980), chap. 1.

26. White, "Contributions of Geographical Analysis," p. 394.

27. Myron Fiering, "Role of Systems Analysis in Water Program Development," *Natural Resources Journal* 16 (1976): 759.

28. White, "Role of Geography," p. 106. See also M. Gordon Wolman, "Selecting Alternatives in Water Resources Planning," *Natural Resources Journal* 16 (1976): 773-791.

29. Frederick L. Strodtbeck and Gilbert F. White, "Attitudes Toward Water," University of Chicago, unpublished manuscript, 1968 (typewritten); Thomas F. Saarinen, "Environmental Perception," in *Perspectives on Environment,* ed. Ian R. Manners and Marvin W. Mikesell (Washington, D.C.: Association of American Geographers, 1974), pp. 252-289; Gilbert F. White, "Public Opinion in Planning Water Development," in *Environmental Quality and Water Development,* ed. Charles R. Goldman (San Francisco: W.H. Freeman & Co., 1973), pp. 157-169; David Lowenthal, ed. *Environmental Perception and Behavior,* Research Papers, no. 109 (Chicago: University of Chicago, Department of Geography, 1967); and the classic, Gilbert F. White, *Choice of Adjustment to Floods,* Research Papers, no. 93 (Chicago: University of Chicago, Department of Geography, 1964).

resource management.

Curiously, there is a significant body of geographical writings devoted to these human issues that receives scant recognition in the decision-making literature. Research classed as cultural ecology, historical geography, and regional geography has had, with few exceptions, little impact on water policy or management practice. Even White states, "Here I confess to a personal bias in favor of research which yields results clearly capable of shaping the direction or quality of public action."[30] As has been belatedly demonstrated in less developed countries, small-scale cultural ecological investigations can not only broaden the range of public choice, but also warn against the human costs of public decisions. Consequently, historical and ex-post project evaluations have been more enthusiastically supported by water managers in recent years.

Substantive critiques of the "decisionistic" framework have alleged an ideological bias in its postulates regarding individual behavior and government policy; a neglect of ethnological and cultural ecological research; and a rigidly inappropriate research methodology.[31] The White school correctly counters that the research design of human geographical studies rarely facilitates direct application of research findings to policy analysis or immediate planning needs, particularly at national or international levels. Although the White school recognizes the potential value of different research methods for identifying cultural factors in hazards adjustment, there remains a clear need for dialogue and synthesis among geographical lines of inquiry.

A more detailed appraisal of the White framework and the geographical research it subsumes, though warranted, would incur a major diversion into hazards research and consequently lies beyond the scope of this monograph. At issue here are the effects of this conceptual framework on the nature and direction of geographical contributions to water resources research. These issues have received little attention and even some abuse in the quest for policy relevance and interdisciplinary coordination.

---

30. White, "Role of Geography," p. 106.

31. Eric Waddell, "The Hazards of Scientism: A Review Article," *Human Ecology* 5 (1977): 69-76; and reply by Gilbert F. White, "Natural Hazards and the Third World--A Reply," *Human Ecology* 6 (1978): 229-231. William I. Torry, "Hazards, Hazes and Holes: A Critique of *The Environment as Hazard* and General Reflections on Disaster Research," *The Canadian Geographer* 23 (1979): 368-383; and reply by Ian Burton, Robert W. Kates and Gilbert F. White, "The Future of Hazard Research: A Reply to William I. Torry," *The Canadian Geographer* 25 (1981): 286-289.

> . . . the distinguishing feature of geographic contributions has been a holistic view . . . It probably continues to be a salient but not unique value of the geographer's participation. Obviously workers in other disciplines have emphasized the same view. In practical terms to bring a geographer into an interdisciplinary investigation often increases the likelihood that the investigation will take a comprehensive view of human and natural phenomena in their interlocking relationships.32

> I am not interested in staking out professional claims in this domain of science. What does seem important is to recognize intellectual problems which call for solutions, and which because of their relation to spatial distribution and human adjustment to differences in physical environment are of interest to geographers.33

More recent review articles by geographers thoroughly inventory the case materials and policy studies alluded to above. They clearly articulate many of the problems water managers face; and yet they tend to overlook or disclaim the conceptual and methodological territory peculiar to geograhical research--territory that perhaps appeared innocently self-evident in the early days of river basin surveys and regional resource development plans. Instead, well-reasoned apologies are offered for the geographer's role in coordinating multidisciplinary research and for probing the general *terrae incognitae* of water resource decision-making.34 These views have acquired casual currency among geographers but have also generated considerable confusion.

> The modern geographer is a mysterious person to one with little formal training in that subject. Geographers seem to be economists, sociologists, or anthropologists wearing false mustaches. At least a reading of this paper helps form a definition of a geographer: some other kind of social scientist--almost any kind--who deals with a particular area. In other words, a geographer is a social scientist in a spatial context.35

Trelease cannot be blamed for his confusion. Water resources geography has shifted from an earlier mainstream concentration on resource estimates and spatial linkages to pioneering studies of perception, hazards adjustment, and efficiency considerations. The benefits from this shift, both to the profession and the public, are well known; its costs in foregone research paths are not so obvious.

---

32. White, "Role of Geography," pp. 116-117.

33. White, "Contributions of Geographical Analysis," p. 393.

34. The tendency toward eclectic sampling (e.g. iceberg harvesting, flood insurance, and systems analysis) in review papers is reflected in Peter Beaumont, "Resource Management: Case Study of Water," *Progress in Physical Geography* 1 (1977): 528-36. K. Smith's review identifies three major problems--efficiency in supply, efficiency in demand, and water quality--on which geographers as well as others are currently focused--see, "Trends in Water Resource Management," *Progress in Physical Geography* 3 (1979): 236-54. Several works have lucidly stressed the importance of merging "physical" and "human" geographical research on water problems, notably Roy C. Ward, "Water: A Geographical Issue," in *Geography: Yesterday and Tomorrow*, ed. E.H. Brown (Oxford: Oxford University Press, 1980), pp. 130-49; and Richard J. Chorley, ed., *Water, Earth and Man* (London: Methuen, 1969).

35. Frank J. Trelease, comment in *Man and Water*, ed. L. Douglas James (Lexington, KY: University of Kentucky Press, 1974), p. 224.

It is ironic that, amidst a burgeoning array of activity in geographical research, "spatial linkages" should be an underdeveloped research theme in water resources geography. With few exceptions the river basin surveys and reclamation reports of forty years ago stand alone as geographical assessments of regional change related to water development. The exceptions include case studies of irrigation systems, interbasin transfers, weather modification, and the impacts of tropical dams. In many cases, however, geographers have focused on a specific water management problem or innovation; assessed its diffusion, areal distribution, or potential applicability; and then moved on to other problems without explicitly stating or critically reviewing what it was they had done. Consequently this process has relied to a large degree on other fields for both its theoretical foundation and problem definition, and as White points out, it has no guarantee of continuation.

Far from simply advocating in revisionist fashion a return to the "good old days" of river basin surveys, I suggest that the problem lies in exploring the geographical character of resource management problems, and in seeking out the conceptual relationships *among* White's six classes of geographical contributions. Expanding the "range of choice" in decision-making, for example, has been a hallmark of water resources geography, but what does it mean, and how is it done?[36]

An answer to this question may be structured using geographical assessments of technological alternatives as an example (figure 2). First, geography is neither a field of technological innovation or technology assessment per se. Geography may inquire, however, into the distribution of prevailing technologies and into problems associated with their occurrence in the landscape. It may then assess the potential applicability and spread of technological innovations in the landscape. By "mapping" the applicability of different choices, the geographer does not necessarily broaden the range of choice, but rather seeks its context in the landscape. Such "mapping" has little accuracy or value if not placed within the broader geographical context of resource distributions, landscape patterns, and social processes (read in White's terminology "resource estimates" and "social guides").

---

36. Perception and choice research in geography has dealt with a broad range of subjects including flood damage reduction, municipal and industrial water supply, water-based recreation, coastal erosion, drought, and water quality. One of the most comprehensive multidisciplinary efforts occurred under White's direction in National Academy of Sciences, *Water and Choice in the Colorado Basin: An Example of Alternatives in Water Management* (Washington D.C.: National Academy of Sciences, 1968).

PROBLEM DEFINITION: Initial identification of problem
    characteristics, appropriate scales of analysis,
    and linkages with other resource management factors.
▼
DESCRIPTION: Survey of the perceived range of choice
    among different social groups; integrated surveys
    of resource distributions, environmental processes,
    and social phenomena affected by the problem (the
    "domain of choice").
▼
ANALYSIS: "Mapping" of resource management alternatives--
    their configuration in the landscape and their effects
    on the domain of choice.
▼
SYNTHESIS: Expansion of the "range of choice" through
    identification of unforeseen or misperceived linkages
    among resource users, use areas, waste flows, and
    institutions.

Fig. 2. Diagram of geographical contributions to
    water resource management

The geographer may participate in literally broadening the range of choice, however, through discovery of unforeseen choices in the course of "mapping" the applicability of previously identified choices. Such discovery originates in the search for linkages among resource supplies, waste flows, uses, and users in the landscape. It may lead to the adaptation of a known management technique to the opportunities and constraints of a different area (read "spatial linkages"). This proposed line of geographical inquiry arrives then, by a circuitous route, directly back at the larger theme of integrated water development.

The Concept of Integrated Development

The concept of integrated water development is broadened in this research from its traditional emphasis on centrally administered river basin development to include the cooperative and creative behavior of water users. Not a type of plan, integrated development is rather a process of adjustment in water use patterns and practices resulting from the search for linkages among water users.[37] The problems of water resources geography in this case become: first, the study of previous adjustments and failures in the landscape; second, the "mapping" of proposed management alternatives; and third, the search for unforeseen linkages among alternatives that are revealed through the mapping process. This does not represent a radical departure from the practice of water resources geography, but rather an attempt to shed some light on the nature of that work, and to make a strong claim for its importance both in the practice of planning and in the actual process of integrated water development.

The following water management advances are offered as examples of integrated water development that complement the innovations generally cited in river basin planning:

1. *Floodplain management*--Adjustment of floodplain occupants to environmental variability represents an extensively developed theme in integrated water management studies in which "structural" and "non-structural" approaches to reducing flood damages are jointly evaluated to guide patterns of land and water use in the floodplains.[38]

---

37. Linkages here refers in part to "spatial linkages" which are established through various forms of interaction and exchange, and in part to relationships governed by allocative rules, administrative practice, and territorial organization.

38. Gilbert F. White, *Choice of Adjustment to Floods,* Research Papers, no. 93 (Chicago: University of Chicago, Department of Geography, 1964); Roy C. Ward, *Floods: A Geographical Perspective* (London: Macmillan Press, 1978); and John L. Sheaffer, *Flood Proofing: An Approach to Flood Damage Reduction,* Research Papers, no. 65 (Chicago: University of Chicago, Department of Geography, 1960).

2. *Integration of irrigation with other agricultural land uses* --Proposals in the 1940's and '50's dealing with the relationship between irrigation and grazing focused on techniques for alleviating fodder shortfalls in dry years and surpluses in wet years, e.g. through joint ownership of irrigated pasture, pasture rental, and marketing of futures feed stocks. More recent arguments have been made for integrating irrigation with wildlife, fisheries, recreation, and settlement schemes.[39]

3. *Wastewater reuse* --By no means a new technique, land application of treated municipal wastewater couples reduced water treatment costs with lower expenditures for agricultural water and fertilizer inputs, thereby offering an alternative to "trade-offs" between urban and rural users. An even more innovative transfer scheme will be practiced by irrigators and the City of Northglenn, Colorado; agricultural water will be treated and used by the city, and then collected and treated before return to the farmers' reservoir during peak irrigation periods.[40]

4. *Conjunctive management of water storage and distribution systems* --Water storage and distribution systems are often developed independently. In addition to the benefits of more closely coordinated project planning, proposals for integrated development have led to: recharge of aquifers with surface water to reduce evaporative losses; mixing of saline and fresh water sources to make use of degraded water supplies; and routing of brackish irrigation return flows through salinity storage sinks.[41]

---

[39]. Jacquelyn L. Beyer, *Integration of Grazing and Crop Agriculture: Resources Management Problems in the Uncompaghre Valley Irrigation Project*, Research Papers, no. 52 (Chicago: University of Chicago, Department of Geography, 1957); Roy E. Huffman, *Irrigation Development and Public Water Policy* (New York: Ronald Press, 1953); R.L. McCown, "The Interaction Between Cultivation and Livestock Production in Semi-Arid Africa," in *Agriculture in Semi-Arid Environments*, ed. Anthony E. Hall et al. (New York: Springer-Verlag, 1979); and Gilbert F. White, *Environmental Effects of Arid Land Irrigation in Developing Countries*, MAB Technical Notes, no. 8 (Paris: UNESCO, 1978).

[40]. James F. Johnson, *Renovated Waste Water*, Research Papers, no. 137 (Chicago: University of Chicago, Department of Geography, 1971); Roger E. Kasperson and Jeanne X. Kasperson, eds. *Water Reuse and the Cities* (Hanover, NH: University of New England Press, 1977); Sheaffer and Roland, Inc., "Northglenn Water Management Program," Report to the U.S. Environmental Protection Agency, (Chicago, 1977).

[41]. Otis W. Templer, "Conjunctive Management of Water Resources in the Context of Texas Water Law," *Water Resources Bulletin* 16 (1980): 309-11; J.R. Teerink, "Conjunctive Use of Ground and Surface Waters," *American Water Works Journal* 60 (1968): 1149-55; Otto J. Helweg, "Regional Ground Water Management," *Ground Water* 16 (1978): 318-21; and Frank Quinn, "Water Transfers: Must the American West be Won Again?" *Geographical Review* 58 (1968): 102-32.

This monograph addresses the problem of "water for Western oil shale development." In conventional water planning this is treated as an ordinary problem in multiple-use, that is, how to produce an optimal quantity of water to satisfy the array of potential uses, and how to effect an optimal redistribution of water supplies to higher value uses in the array. Our objectives in this study are closely related; but rather than focus on the volumes of water to be transferred or developed, we shall investigate the variety of potential linkages among industrial and agricultural users in the case study area.

The simplest type of exchange would appear at first glance to be direct water rights sales. A more creative proposal for integrated development has been suggested, however, in which oil shale water demands are met in part by financing improvements in the efficiency of agricultural water use. Transfer of water that is "salvaged" (through improvements in irrigation efficiency such as canal linings, conversion to drip irrigation, or computerized irrigation scheduing) to higher value uses would result both in conservation of existing water supplies and reductions in agricultural water pollution. The potential significance of agricultural water conservation in water resources management is readily made apparent by the fact that over 93 percent of all water withdrawn in the Upper Basin is for irrigation, and that, of this, roughly 30 percent is productively consumed through evapotranspiration while a full 70 percent is lost in conveyance, inefficient distribution, operational spills, and salt leaching.[42] Most conservation practices reduce these non-consumptive uses, though smaller gains are also sought in reduced evaporation and crop water use. The fact that water not consumed by the crop returns to a stream for subsequent reuse, albeit often diminished in quality or quantity, raises institutional barriers to any proposal that might increase or shift the location of water use. Our research problem thus becomes one of identifying the range of geographical conditions under which such conservation exchanges could take place in western Colorado.

A rather circuitous path has been laid out in order to establish the nature and importance of geographical research in water resources management. At the risk of exacerbating semantic confusion among practitioners of integrated river basin development, the concept was recast in more general terms as a process of

---

42. U.S. Water Resources Council, *The Nation's Water Resources 1975-2000*, vol. 4: *Upper Colorado Region* (Washington, D.C., 1978), p. 17.

adjustment in water use or practice resulting from the search for cooperative linkages among water users. Gilbert White's classification of geographical research in water resources provided a basis for more formally developing the geographical dimensions of integrated water development, as well as for beginning to seek out the conceptual unity of elements in that classification. Chapter three begins the demonstration of this geographical approach by outlining management arenas and water control points in the Colorado River Basin that shape local water practice in western Colorado.

CHAPTER III

A GEOGRAPHY OF WATER CONTROL ON THE COLORADO RIVER IN WESTERN COLORADO

The dynamic nature of hydrologic phenomena and of human efforts to modify them make the delimitation of local water problems particularly difficult. Even clear physiographic boundaries between contiguous drainages become permeable as a result of transbasin diversions and sometimes groundwater pumping or flow paths. Within any given basin, upstream water withdrawals affect downstream uses. Institutional arrangements in the form of compacts, water rights, treaties, and water quality regulations have been formulated to manage these effects with the result that downstream uses may also affect upstream uses. Thus, water use in any given location often reflects strategic objectives with respect to water uses in other locations.

Because rules developed for larger arenas, such as the international basin or the entire Colorado River Basin, tend to dominate those of more local arenas, water control may be initially understood as a "nested hierarchy" of institutions corresponding to specific administrative areas and administered at specific control points in the hydrologic system. In this chapter the structure and uncertainities of water control on the Colorado River are surveyed for their effects on water use and water development strategies in the two-county study area in western Colorado.

There is a touch of pride voiced in the common remark that the Colorado is "the most developed and litigated" river in the world, "the most cussed and discussed." The scarcity theme underlying these claims comes out most forcefully in issues affecting the entire basin or rather in those affecting the Lower Basin, Mexico, and transbasin diverters, for those are the areas of acute water competition and conflict. As will become evident, the Upper Basin's "problems" have historically stemmed from having, first, a perceived water surplus and, second, a set of grand expectations for future resource development.

Attention is directed first to international agreements with Mexico which affect water rights in the entire basin, then to the apportionment of water between the Upper and Lower Basins, within the Upper Basin, the State of Colorado, Colorado Water Division 5, and finally, within the study area of two counties in western Colorado.

International Agreements

Prior to the 1930's, disputes with Mexico over the Colorado River centered on problems of flood control, boundaries, and shifting channel locations. Plans for irrigation projects in the Colorado River delta and the Mexicali Valley followed shortly after completion of Boulder Dam, the All-American Canal, and extensive reclamation projects. Increases in consumptive losses from these projects initiated conflict over the supply of water available to each country. An international apportionment of water was agreed to in 1944 after considerable negotiation with Mexico and among the seven basin States--and before the discovery of several serious errors in assumptions on the American side (figure 3). The treaty of 1944 apportioned the Colorado as follows:

1. Guaranteed annual delivery of 1.5 million acre-feet per year (MAF)
2. In years of surplus to increase the amount delivered to Mexico up to a maximum of 1.7 MAF, and in years of drought to curtail the amount delivered to Mexico in proportion to the reduction in consumptive use in the United States.[1]

Errors included first, the assumption that average annual virgin flows of the Colorado River exceed 17 MAF, a figure based on an abnormally wet period from 1921 to 1942, and second the assumption that treaty deliveries "from any source or origin" contained no implied stipulations regarding water quality, or lack thereof.

Conflict over river salinity and American treaty assumptions escalated with the onset of highly saline discharges from the Welton-Mohawk Drain in Arizona, the return to lower average annual flows, and the growth of Mexican reclamation activity. As Myron Holburt pointed out, the salinity issue involved not simply a Federal treaty, but rather the full body of institutions controlling water use in both Upper and Lower basins.[2] An agreement reached in

---

1. U.S. Department of State, Treaty Series 994, Article 10; delivery locations are specified in Article 11.

2. Myron B. Holburt, "International Problems," in *Values and Choices in the*

Fig. 3. Map of water development in the Colorado River Basin

1973 by the International Boundary and Water Commission stipulated the following standards and measures:

    a) . . . the approximatedly 1,360,000 acre-feet (1,677,545,000 cubic meters) delivered to Mexico upstream of Morelos Dam, have an annual average salinity of no more than 115 p.p.m ± 30 p.p.m. U.S. count (121 p.p.m. ± 30 p.p.m. Mexican count) over the annual average salinity of Colorado River waters which arrive at Imperial Dam......

    b) The United States will continue to deliver to Mexico on the land boundary at San Luis and in the limitrophe section of the Colorado River downstream from Morelos Dam approximately 140,000 acre-feet (172,869,000 cubic meters) annually with a salinity substantially the same as that of the waters customarily delivered there.

    c) . . . each country shall limit pumping of groundwater in its territory within five miles (eight kilometers) of the Arizona-Sonora boundary near San Luis to 160,000 acre-feet (197,358,000 cubic meters) annually.

    d) Completion of salinity control works in the United States and assistance to Mexico in obtaining financing and planning for such works in Mexico.[3]

Implementation of this agreement was assigned to the Secretary of the Interior in the Colorado River Basin Salinity Control Act of 1974.[4] The Act established a set of salinity control objectives and measurement points both upstream and downstream of Imperial Dam. Title II laid out a salinity control program for basin areas upstream of Imperial Dam, in essence the Colorado River Water Quality Improvment Program (CRWQIP) discussed later in this chapter.

Basin-Wide Apportionment

    . . . to provide for the equitable division and apportionment of the use of the waters of the Colorado River System; to establish the relative importance of different beneficial uses of water; to promote interstate comity; to remove causes of present and future controversies; and to secure the expeditious agricultural and industrial development of the Colorado River Basin, the storage of its waters, and the protection of life and property from floods.[5]

Implicit in these words is a hard-won and cautious cooperation among the basin States, in part motivated by a desire to lessen the intense upstream-downstream controversies and strategic behavior that had impeded political collaboration to obtain major Federal public works projects. The classic upstream problem in appropriation states is securing adequate title to future water needs, while the classic downstream problem is protection against disenfranchisement by such future upstream development. The compact formalized the administrative division into Upper and Lower Basins at Lees Ferry on the Colorado River; it resolved several major

---

Development of the Colorado River Basin, ed. Dean F. Peterson and A. Berry Crawford, (Tucson: University of Arizona Press, 1977), pp. 220-238.

    3. Minute 242, International Boundary and Water Commission, United States and Mexico, August 30, 1973. TIAS 7708, 59 Stat. 1219.

    4. P. L. 93-320, 88 Stat. 266, 43 U.S.C. 1571.

    5. Colorado River Compact, Art. I. Signed November 24, 1922; approved in the Boulder Canyon Project Act of 1928; ratification proclaimed on June 25, 1929. 46 Stat. 3000.

questions, and created others--the most pressing question being the actual volume of water apportioned to the Upper and Lower Basins.

> Art. III (a) There is hereby apportioned from the Colorado River System in perpetuity to the Upper Basin and to the Lower Basin, respectively, the exclusive beneficial consumptive use of 7,500,000 acre-feet of water per annum, which shall include all water necessary for the supply of any rights which may now exist . . .
>
> (c) If, as a matter of international comity, the United States of America shall hereafter recognize in the United States of Mexico any right to the use of any waters of the Colorado River system, such waters shall be supplied first from the waters which are surplus over and above the aggregate of the quantities specified in paragraphs (a) and (b); and if such surplus shall prove insufficient for this purpose, then, the burden of such deficiency shall be equally borne by the Upper Basin and the Lower Basin, and wherever necessary the States of the Upper Division shall deliver at Lee Ferry water to supply one-half of the deficiency so recognized in addition to that provided in paragraph (d).
>
> (d) The States of the Upper Division will not cause the flow of the river at Lee Ferry to be depleted below an aggregate of 75,000,000 acre-feet for any period of 10 consecutive years reckoned in the continuing progressive series beginning with the first day of October next succeeding the ratification of this compact . . .
>
> (f) Further equitable apportionment of the beneficial uses of the waters of the Colorado River system unapportioned by paragraphs (a), (b), and (c) may be made in the manner provided in paragraph (g) at any time after October 1, 1963, if and when either basin shall have reached its total beneficial consumptive use as set out in paragraphs (a) and (b).[6]

A serious discrepancy arose between III(a) and III(d) above as it became clear that the average annual virgin flow at Lees Ferry falls well below the 15 million acre-feet (MAF) necessary to fill both provisions, much less the amount required to satisfy the treaty with Mexico. Adjusted estimates of annual virgin flow fall as low as 13.5 MAF reflecting both tree-ring data and more recent periods of record.[7] Article III(a) implies a 50-50 division of available water while III(d) calls for delivery of a set volume of water to the Lower Basin. Naturally, the Lower Basin States regard III(d) as binding while the Upper Basin States argue that III(d) was based on a mutual misunderstanding and should give way to III(a). A related problem then attaches to Art. III(c), referred to as the "Mexico burden." Upper Basin States contend that because the lower States refuse to include their tributary flows (e.g. the Gila River) as part of total basin yield, adhere to an inequitable interpretation of the compact, and even then consume in excess of their compact apportioned water (including some of the "surplus" from undeveloped Upper Basin waters), that the entire Mexico burden should fall on the Lower Basin. The effects of these disputes on water management in the Upper Basin have been first, to increase the uncertainty regarding water availability, and second, to reinforce the

---

6. Ibid., Art. III.

7. Colorado Department of Natural Resources, *13(a) Assessment* (1979); Office of Technology Assessment, *An Assessment of Oil Shale Technologies* (Washington, D.C., 1980); and Musick and Cope, "Briefing Paper: Water for Western Energy Development," unpublished (Boulder, CO, 1981).

persistent strategy of fully developing compact waters so as to avoid losing them through reapportionment, water quality legislation, or other means.

Basin Development Acts

In spite of the disputes regarding compact provisions, the compact facilitated collaboration on large-scale multipurpose water projects. The Boulder Canyon Project Act of 1928 was the first of these, authorizing the All-American Canal and Hoover Dam for purposes of flood control, navigation, reclamation, and other beneficial uses.[8] It should be pointed out that the Boulder Canyon Project Act enabled final ratification of the Colorado River Compact of 1922 by stipulating conditions under which only five or more States would be required to approve the compact; the Act also congressionally apportioned water among the Lower Basin States in a manner later upheld by the U.S. Supreme Court in *Arizona v. California*.[9] The Boulder Canyon Project Adjustment Act of 1940 then established a Colorado River Development Fund to come from receipts of power sales at Hoover Dam for the purpose of conducting water development studies and constructing projects "equally distributed among the states of the upper division and the states of the lower division."[10] This would become an important impetus for project development in the Upper Basin.

Storage works in the Upper Basin have served the dual purposes of enabling those States to meet compact delivery requirements at Lees Ferry as well as to stimulate water development for local needs. The Colorado River Storage Project Act of 1954, the foremost vehicle of storage development in the Upper Basin, authorized the following: construction of four major reservoir projects (Curecanti, Flaming Gorge, Navajo, and Glen Canyon); construction of sixteen "participating projects" (including the Silt Project in the study area); and investigation of nineteen potential participating projects (including the West Divide, Bluestone, Battlement Mesa, and Grand Mesa projects--in or near the study area).[11] Power revenues credited to an Upper Colorado River Basin Fund aid in defraying the operation and maintenance costs; excess funds are apportioned to the Upper Basin States for repayment of participating project construction costs. In this way some of the benefits of storage for

---

8. 45 Stat. 1057, 43 U.S.C. 617.

9. 373 U.S. 546, 83 S. Ct. 1468, 10 L. Ed. 2d 572 (1963).

10. Sec 2(d), 54 Stat. 774; as amended in 621 Stat. 284.

11. 70 Stat. 105; as amended in 76 Stat. 102 and 78 Stat. 852; 43 U.S.C. 620.

the Lower Basin have been redistributed to developing areas in the Upper Basin.

Comprehensive basin development planning was again taken up in 1968 with the Colorado River Basin Project Act which authorized a seventeen state western water study.[12] The Act established several important controls on water development in the Colorado River Basin: the Treaty with Mexico was affirmed as a " . . . national obligation . . . , the first obligation of any water augmentation project planned pursuant to [this Act] . . . ;"[13] a ten year moratorium was placed on Federal investigations of water importation into the Colorado River Basin;[14] and the Secretary of the Interior was required to promulgate operating criteria for major main stem reservoirs. Operating criteria put forward in 1970 calling for a minimum annual release of 8.23 MAF from Lake Powell would appear to favor a Lower Basin interpretation of the Colorado River Compact, though this has not been conceded by the Upper Basin States.[15] The Act and operating criteria also imply future curtailment of Lower Basin water use, however, including that of the newly constructed Central Arizona Project. Curtailment of Lower Basin uses will probably accompany Upper Basin development in the absence of augmentation plans for the basin, that is, without further groundwater development or water importation into the basin.

Basin-Wide Water Quality Planning

Interstate consideration of Colorado River water quality problems began in 1960 and continued in the form of technical conferences through 1972. In 1972 the Bureau of Reclamation put forward a basin-wide technical study known as the Colorado River Water Quality Improvement Program (CRWQIP).[16] This study was later incorporated as Title II of the Colorado River Basin Salinity Control Act of 1974 which authorized construction of four salinity control projects (including the Grand Valley program of canal lining and on-farm improvements in western Colorado) as well as feasibility assessments for twelve additional projects in the basin. The

---

12. 82 Stat. 885, 43 U.S.C.A. 1501 et seq.

13. Ibid., 1512.

14. The 1968 moratorium was extended for an additional ten years in the Reclamation and Safety of Dams Act of 1978, P. L. 95-578, 95 Stat. 2472.

15. U.S. Department of the Interior, "Criteria for Coordinated, Long-Range Operation of the Colorado River Reservoirs," *Federal Register* 35 (1970): 8951-2.

16. U.S. Department of the Interior, Bureau of Reclamation "Colorado River Water Quality Improvment Program," special report (1972); see also U.S. Environmental Protection Agency, "The Mineral Quality Problem in the Colorado River Basin," *Conference in the Matter of Pollution of the Interstate Waters of the Colorado River and Its Tributaries, Seventh Session,* vol. 1, (1972).

legislation explicitly sought not to alter existing water rights allocations or compact agreements, following instead the reasoning that the national obligation to Mexico should be satisfied principaly through public expenditure on structural salinity control projects. One quarter of the costs are to be borne by the Upper and Lower Basin Development Funds with no more than 15 percent of this sum to be paid by the Upper Basin Fund. As a result, the costs of projects located in the Upper Basin fall to a large extent on the Lower Basin users who presumably will benefit most from the legislation.[17]

The 1972 technical conference sidestepped the issue of numerical criteria for implementation of the salinity control program, but with passage of the Federal Water Pollution Control Act Amendments of 1972, the Environmental Protection Agency (EPA) required that each state adopt such criteria by 1975. The Colorado River Basin Salinity Control Forum, organized in anticipation of the Salinity Control Act, developed numerical criteria for the entire basin, to be measured at three locations.

Location	Flow-Weighted Annual Average Total Dissolved Solids
Hoover Dam	723 mg/l
Parker Dam	747 mg/l
Imperial Dam	879 mg/l

Each State subsequently adopted these criteria, and EPA approved them in 1976.[18] The downstream location of these control points meant that they would serve more to monitor the program than to enforce water quality standards at pollution sources. The Environmental Defense Fund unsuccessfully challenged EPA on this open-ended approach, calling instead for enforcement standards at State boundaries on all major streams.[19] Upper Basin States avoided any such plan of implementation that might limit full development of

---

17. Title II, 88 Stat. 266, 43 U.S.C. 1571.

18. Colorado River Water Quality Control Commission, Regulation 3.9.1 et seq. "Colorado River Salinity Standards" and "Plan of Implementation," as amended (May 5, 1980).

19. *Environmental Defense Fund v. Douglas M. Costle, et al.,* "Appellants' Petition for Rehearing and Suggestion for Rehearing en Banc," U.S. Court of Appeals, District of Columbia Circuit (1981).

their compact apportioned waters, and indeed they had actively participated in the Salinity Forum in part to insure that water quality legislation would not be employed to those ends. The initial plan of implementation in Colorado, based on the National Pollutant Discharge Elimination System (NPDES) and zero-salt discharge for all new industrial sources, was later repealed and replaced with a regulation including salinity in the more general "208" area-wide water quality planning process.[20]

National Water Programs and Policies

Many Federal programs influence water use at the local level either directly or indirectly. Rather than embark upon a wide-ranging inventory of these, several of the more important regulations for this case study are identified.

1. "208" area-wide water quality planning, mandated in section 209 of the Clean Water Act of 1977 and administered by the Colorado West Area Council of Governments.[21]
2. "404" dredge and fill permits, also of the Clean Water Act and administered by the Army Corps of Engineers for any proposed action in a navigable waterway, have become increasingly controversial in the Gunnison and other tributary basins.[22]
3. The Fish and Wildlife Coordination Act of 1958 requiring consideration of wildlife issues in water resource development programs and mitigation of development impacts.[23]
4. The Endangered Species Act of 1973 has been invoked in response to dam proposals on the Colorado River main stem in western Colorado wherein two nationally endangered species--the humpback chub and the Colorado squawfish--and four species on the Colorado endangered species list have been found. Designation of the humpback chub and squawfish as endangered species was recently revoked.[24]

---

[20]. Cf. Code of Colorado Regulations 3.10.0 et seq. (1976) and (1981).

[21]. Clean Water Act of 1977, P. L. 92-500, 33 U.S.C. 1251. Colorado West Area Council of Governments, *Colorado West Area 208 Plan, Final Main Report* (Rifle, CO., 1979) and update memorandum (1980).

[22]. Clean Water Act of 1977, Sec. 404; "Corps' Handling of Water Act Lashed," *Denver Post* (June 14, 1981).

[23]. Fish and Wildlife Coordination Act of 1958, 72 Stat. 563, 16 U.S.C. 661.

[24]. Endangered Species Act of 1973, 87 Stat. 884, 16 U.S.C.A. 1531 et seq.; U.S. Department of the Interior, Bureau of Land Management, *Final Environmental Impact Statement: Proposed Development of Oil Shale Reserves by the Colony Development Operation in Colorado*, vol. 1. (n.d.); and *Colorado River Water Conservation District et al. v. Cecil D. Andrus et al.* (August 1981).

5. The National Environmental Policy Act of 1969 requiring impact assessments for all Federal actions has led to the preparation of several voluminous oil shale and coal development impact assessments in which water resources were explicitly considered.[25]

6. Economic feasibility requirements, first stipulated in the Flood Control Act of 1936 and the Reclamation Project Act of 1939, have impeded final authorization of several "participating projects" including the West Divide Project located in the study area.[26]

In addition, unresolved Federal water policies have generated considerable uncertainty over the security of water rights in the Colorado River Basin. Federal reserved rights for management of Bureau of Land Management, Forest Service, and Naval Oil Shale Reserves lands have not been quantified but have been made subject to jurisdiction of the State courts.[27] In 1979 Solicitor Krulitz also claimed "non-reserved" Federal water rights for purposes not originally specified in the reservation of public lands--a claim that was surrounded by controversy until its recent revocation.[28] The States have faced a similar set of issues in attempting to establish minimum instream flow reservations. Large-scale manipulation of climactic patterns, e.g. through weather modification experiments, could cause future adjustments in water rights systems, though this does not seem imminent in the study area.[29] Finally, quantification of Indian reserved rights in the White River and San Juan River Basins could affect water rights and compacts throughout the Colorado River Basin. The Winters Doctrine

---

25. National Environmental Policy Act of 1969, P. L. 91-190, 42 U.S.C.A. 4321 et seq.

26. Flood Control Act of 1936, 49 Stat. 1570, 33 U.S.C. 701. Reclamation Project Act of 1939, 53 Stat. 1187, 43 U.S.C. 485. Modification of projects to include energy water uses has improved their overall feasibility but still often fails to result in feasible irrigation development, cf. U.S. Bureau of Reclamation, *Potential Modification in Eight Proposed Western Colorado Projects for Future Energy Development* (Salt Lake City: Upper Colorado Region, 1980), and "Water Board Discusses West Divide Storage Project Future," *Rifle Telegram* (October 28, 1981).

27. *Colorado River Water Conservation District v. United States* 424 U.S. 800, 96 S. Ct. 1236, 47 L. Ed. 2d 483; and *United States v. District Court for Water Division 5*, 401 U.S. 527, 91 S. Ct. 1003, 28 L. Ed. 2d 284. Recent assessments have downplayed the impacts of Federal reserved rights, see Frank J. Trelease, "Federal-State Problems in Packaging Water Rights," in *Water Acquisition for Mineral Development* (Boulder, CO: Rocky Mountain Mineral Law Foundation, 1978); and idem, "Federal Reserved Water Rights since PLLRC," 54 *Denver Law Journal* (1977): 473-492.

28. Krulitz Opinion No. M-36914, June 25, 1979; amended by Solicitor Martz, Jan. 16, 1981; later revoked, "Watt, in Water Policy Shift, Says U.S. Cannot Pre-empt State Right," *New York Times* (September 12, 1981).

29. Edward A. Morris, "Institutional Adjustment to an Emerging Technology: Legal Aspects of Weather Modification," in *Human Dimension of Weather Modification,* ed. W.R. Derrick Sewell, Research Papers, no. 105 (Chicago: University of Chicago, Department of Geography, 1966), pp. 179-88.

implies that the Indian reserved rights priorities date at least to the founding of the reservation, making Indian rights senior to most other rights in the Basin. Many issues concerning administration, permissible uses, and transfers of Indian reserved rights remain unresolved.30

Omitted from this discussion of Federal actions are the ubiquitous construction grants programs, energy facility siting requirements, and technical water planning assistance programs administered by various agencies. Those related to water conservation will be treated in greater detail in subsequent chapters.

Upper Basin Apportionment

Whereas the Lower Basin States were unable to agree to a division of water among themselves, the Upper Basin States saw such agreement as a prerequisite to developing their compact apportioned water supplies.

> The formulation of an ultimate plan of river development, therefore, will require selection from among possibilities for expanding existing or authorized projects as well as from the potential new projects. Before such a selection of projects can be made, it will be necessary that the seven Colorado River Basin States agree upon their respective rights to deplete the water supply of the Colorado River or that the courts apportion available water to them.31

> Further development of the water resources of the Colorado River Basin, particularly large-scale development, is seriously handicapped, if not barred, by lack of a delimitation of the rights of individual states to utilize the waters of the Colorado River system.32

The Upper Colorado River Basin Compact of 1948 allocated 50,000 acre-feet to Arizona, and apportioned in perpetuity the remaining waters on the following basis (Art. III(a)):33

Colorado	51.75%
New Mexico	11.25%
Utah	23.00%
Wyoming	14.00%

This formula was based on an consensual balance of many criteria including basin yield, existing uses, potential use, and stream channel losses; it is only to be invoked, however, in the event of failure to meet delivery requirements to the Lower Basin.

---

30. *Winters v. United States,* 207 U.S. 564, 28 S. Ct. 207, 52 L. Ed. 340 (1908).

31. U.S. Department of the Interior, *The Colorado River: A Natural Menace Becomes a National Resource,* H.R. Doc. No. 419, 80th Cong., 1st Sess. (1947), p. 13.

32. Ibid., letter of transmittal from Secretary Krug, July 19, 1947.

33. 63 Stat. 31, not codified.

The Compact assigned reservoir losses to the states in which they occur (Art. V) and established an interstate administrative agency, the Upper Colorado River Commission, to facilitate its implementation and to serve as a forum for policy matters and technical issues (Art. VIII). Articles eleven through fourteen further apportion waters in the following tributary drainages: the Little Snake River, the Yampa River, the Green River in Wyoming and Utah, and the San Juan River. Of these, only the Yampa River agreement might bear on water use in west-central Colorado, and then only if transbasin diversions are contemplated. The fact that no compact agreements exist for the Colorado River main stem or White River could later lead to conflict over water appropriation for energy development in those drainages.

Water and Energy in the Upper Basin

Recent assessments of Upper Basin water supplies stand in contrast with studies of the entire Colorado River Basin. They have generally affirmed the adequacy of state compact-apportioned waters to meet existing uses and projected water demands, assuming moderate energy development scenarios (figure 4).

> It is estimated that the water demands of an EET [emerging energy technology] industry of 1.5 million bbl/day, as well as the water demands of the associated growth, could be satisfied from surface supplies without having to significantly reduce (if at all) other projected consumptive uses in the Upper Basin.[34]

> . . . if the reservoirs and the pipelines are built, flows do not decrease [below 13.8 MAF per year], and [non-energy water demand in] the region develops at a medium rate, there should be sufficient surplus water to support an industry of over 2 million bbl/day through 2000.[35]

These studies share several important assumptions: continued low virgin flows (13.8 MAF); Upper Basin obligations to deliver 7.5 MAF per year to the Lower Basin and half of the Mexico burden (.75 MAF excluding losses); construction of additional pipelines and reservoirs; growth in all non-energy water demands including Federal irrigation projects; and a wide range of energy water use technologies. The Colorado study also assumes purchase of all existing Federal reservoir water.

These studies imply that a surplus exists in the Upper Basin that has been overshadowed by the water scarcity myth on the Colorado River. In line with the "use it or lose it" perspective of the prior appropriation doctrine, and in spite of the apparent stability of the Colorado River compact, this "surplus" is regarded as a "problem" that urges on full development of the States' compact

---

34. Colorado Department of Natural Resources, *13(a) Assessment,* p. 1-1.

35. Office of Technology Assessment, *An Assessment of Oil Shale Technologies,* p. 359.

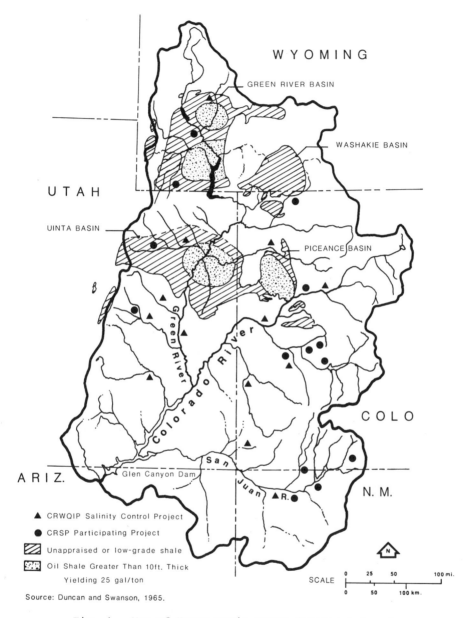

Fig. 4. Map of Upper Basin water development and oil shale resources

waters to prevent their being usurped in some manner by Lower Basin users or transmountain diverters. Failure to curtail Lower Basin users to their compact limitations and inclusion of evaporative losses in the Upper Basin share of the Mexico Burden, the worst case yet considered, could result in Upper Basin deficits of as much as 1.6 MAF by the year 2000.[36] Transbasin water diversions have climbed steadily over the past three decades, though with conflicts and delays in new projects, diversions have levelled off during the past decade. Related threats are perceived to come from water quality legislation aimed at salinity control, fish and wildlife protection, and pollution control.[37]

Upper Basin water studies rarely consider the effects of implementing the Upper Colorado River Compact on individual tributaries in the event of drought; nor do they fully incorporate water transfer restrictions or environmental constraints in their models of trade-offs among higher and lower value uses.[38] Such issues require a more detailed treatment than is generally conducted at the river basin scale. The important point is that at the regional scale, a perceived water surplus, and not scarcity, has become the chief guide for policy formulation in the Upper Basin. Extrabasin threats to the "paper surplus" and the prospect of increasing water appropriations in the Upper Basin have fostered policies that store, develop, and consume water first--and consider broader issues later.

State Water Management

The right to appropriate water in Colorado is solidly embedded in Article XVI of the State Constitution of 1876:

> The water of every natural stream, not heretofore appropriated, within the State of Colorado, is hereby declared to be property of the public, and the same is dedicated to the use of the people of the State, subject to appropriation as hereinafter provided [section 5].
>
> *Priority of appropriation* shall give the better right as between those using the water for the same purpose...... The right to divert and put water *to a beneficial use shall never be denied*. [section 6, emphasis added].

---

36. Musick and Cope, "Briefing Paper for Western Energy Development," unpublished (Boulder, CO, 1981).

37. David L. Harrison, "Federal Regulation of Appropriations of Water in the Name of Protecting Water Quality," in *Proceedings*: *Water Resource Allocation Laws and Emerging Issues* (Boulder, CO: University of Colorado School of Law, 1981).

38. These limitations are explicitly recognized in Colorado Department of Natural Resources, *13(a) Assessment* (1979). Transfer restrictions were found to have significant impacts on water availability in A.B. Bishop and R. Narayanan, "Competition of Energy for Agricultural Water Use," *Journal of the Irrigation and Drainage Division ASCE* 105 (1979): 372-35; and Constance M. Boris and John Krutilla, *Water Rights and Energy Development in the Yellowstone River Basin*: *An Integrated Analysis* (Baltimore: Johns Hopkins University Press for Resources for the Future, 1980). Tributary relationships were considered in the Colorado Department of Natural Resources, *13(a) Assessment* (1979).

In 1969 "waters of the state" were redefined to include all surface water and groundwater tributary to natural surface streams, alleviating the common conflicts between groundwater and surface water development.[39] The appropriation doctrine, often simply characterized as "first in time, first in right," embodies a rule of property based on first occupancy, whereby rights are satisfied in order of seniority. It varies in important ways from State to State as to the nature and requirements of a water right. In Colorado, a right is acquired upon demonstration of an "intent" to appropriate, and second, by actually putting the water to a "beneficial use;" though the precise meaning of these terms has been interpreted at length and in the end left purposefully underdefined.

> "Beneficial use" is the use of that amount of water that is reasonable and appropriate under reasonably efficient practices to accomplish without waste the purpose for which the appropriation is lawfully made and, without limiting the generality of the foregoing, includes the impoundment of water for recreational purposes, including fishery or wildlife. For the benefit and enjoyment of present and future generations, "beneficial use" shall also include the appropriation by the State of Colorado, in the manner prescribed by law, of such minimum flows between specific points or levels for and on natural streams and lakes as are required to preserve the natural environment to a reasonable degree.[40]

The longstanding requirement of diverting water to establish beneficial use was omitted from the 1973 revised statutes in recognition of instream flow uses.[41] Modification of a water right through reuse, exchange, transfer, conservation, or augmentation entails specific statutory proscriptions and administrative requirements that will be taken up in detail in subsequent chapters. Water rights perfected through beneficial use become subject to protection and enforcement in accordance with the prior appropriation doctrine.

Although Colorado does not require permits for appropriation--as do do most other appropriation States--enforcement and protection of water rights is predicated on compliance with

---

39. C.R.S. 1973, 37-92-102(1) and 37-92-103(13). Non-tributary ground water came under State administration in the 1965 Ground Water Management Act, C.R.S. 1973, 37-91-101 et seq. There is a presumption that ground water is tributary unless shown to be otherwise. See David L. Harrison and G. Sandstrom Jr., "The Groundwater-Surface Water Conflict and Recent Colorado Legislation," *University of Colorado Law Review* 43 (1971): 1-48.

40. C.R.S. 1973, 37-92-103. The exact nature of beneficial use is site and situation specific, see *City and County of Denver v. Sheriff,* 105 Colo. 193, 96 P.2d 836 (1940). Running excess water in ditches and failing to maintain ditches are prohibited in C.R.S. 1973, 37-84-107, 108; but otherwise "waste" is loosely defined. Speculation was also prohibited, see *Colorado River Water Conservation District v. Vidler Tunnel Co.* 197 Colo. 413, 594 P.2d 566 (1979), yet conditional rights may be acquired upon demonstration of an "intent" to develop the water at a future date (C.R.S. 1973, 37-92-103). Demonstration of "reasonable diligence" is required to maintain a conditional right which, upon perfection, obtains a priority dating back to its original filing by the postponement or "relation back" doctrine.

41. C.R.S. 1973, 148-121-3(7). *Empire Water & Power Co. v. Colorado Water Conservation Board* 197 Colo. 469 (1979) upheld the constitutionality of such appropriations; but administrative rules are still in flux, C.R.S. 1973, 37-92-102(3), and Colorado Senate Bill No. 414 (1981).

certain administrative procedures, the first of which is a court adjudication of water title, a practice initiated in the Irrigation Law of 1879. The priority of a right is based first upon the year of adjudication; rights adjudicated in the same year are then ranked by date of appropriation, making unadjudicated rights technically unenforceable.[42] The 1879 law also established the State Engineer's Office which divided the State into seventy irrigation districts. District boundaries generally conformed to surface drainage basins, and their size was based on the current level of irrigation activity. One water commissioner had to administer all of the decrees in one irrigation district. Upstream-downstream conflicts among water users in different districts led to the grouping of districts into major basin-wide divisions in the 1887 Irrigation Law (figure 5).

The duties of State officials and water rights holders shifted gradually between the 1905 irrigation law and the 1969 Water Rights Determination and Administration Act (C.R.S., 1973, 37-92-101 et seq.). Under the 1969 Act seven major water divisions were established containing 72 water districts (C.R.S., 1973, 37-92-201 et seq.). Each division has a water court consisting of a water judge, clerk, and referees; it also has a division engineer, his water commissioners, and their deputies. Adjudication proceedings were made continuous rather than annual or irregular, and a comprehensive tabulation of all conditional and absolute decrees was required of the Colorado Division of Water Resources. More detailed discussion of this tabulation is reserved for the following chapters in which it is used extensively in the case study analysis. Conditional water rights holders are required to file for findings of due diligence, and all water rights holders are to file timely statements of opposition to any disputed entries in the water rights tabulation. Trelease interprets these more formal procedural requirements as a shift toward the regulatory system found in "permit" States.[43]

Statewide water policy has evolved in response to the emergence and evolution of public interests in water resource management. In 1903 Elwood Mead called specifically for public interests to be more fully represented in Colorado water rights proceedings. Over the past several decades the State has created the following executive departments and commissions charged with the

---

42. Elwood Mead, *Irrigation Institutions* (New York: The Macmillan Co., 1903) p. 147.

43. Frank L. Trelease, *Water Law,* 3rd ed. (St. Paul: West Publishing, 1979), p. 142.

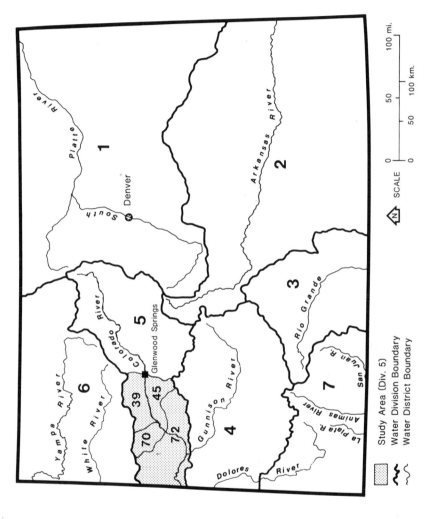

Fig. 5. Map of water administration in Colorado

formulation of water policy and implementation of legislated programs:
1. Colorado Water Conservation Board (C.R.S., 1973, 37-60-101)
2. Colorado Groundwater Commission (C.R.S., 1973, 37-90-104)
3. Colorado Water Quality Commission (C.R.S., 1973, 25-8-201)
4. Colorado Water Resources and Power Development Authority (C.R.S., 1981 supp., 37-95-101).

Creation of the Water Resources and Power Development Authority follows the very recent shift in Federal-State relations toward increased participation in water project financing.

Water Control in Division 5

The Upper Colorado River main stem and the study area of this research fall entirely within Colorado Water Division 5 (figure 6). Any discussion of "control points" in the Upper Colorado must first consider linkages with other basins, such as the South Platte and Arkansas drainages in which Front Range metropolitan growth is concentrated. Transbasin diversions from the Upper Basin have signficantly affected water development strategies in Western Colorado.

Colorado water law gave early recognition to the right to appropriate water from one basin for use in another.[44] Appropriation of Colorado River headwaters for use on the Front Range began in 1880 with the Grand River Ditch. Longstanding conditional decrees were maintained by Front Range water organizations as part of what Corbridge has termed the "great and growing cities" doctrine.[45] Fears of future exhaustion of the State's share of Colorado River Compact waters led to conflicts between eastern and western Colorado water interests culminating in the massive Colorado-Big Thompson (C-BT) and Blue River diversion projects. U.S. Senate Document 80 framed out a compromise plan for the C-BT project that included replacement storage for transbasin diversions in Green Mountain Reservoir.[46] Downstream "calls" on the Colorado River thus lead first to the release of replacement water from Green Mountain Reservoir, up to a total volume of 50,000 acre-feet, and then subsequently to the curtailment of junior appropriators.[47] As is

---

44. *Coffin v. Left Hand Ditch Co.* 6 Colo. 443 (1882).

45. James N. Corbridge, Jr. "Outline--Application of the Law of Prior Appropriation," in *Water Resources Allocation*: *Law and Emerging Issues* (Boulder, CO: University of Colorado School of Law, 1981).

46. U.S. Senate Doc't 80, "Synopsis of Report on Colorado-Big Thompson Project," 1937.

47. The expression "call on the River" refers to an announcement by the Division Engineer that senior water rights are not being met and that regulation of diversions will begin.

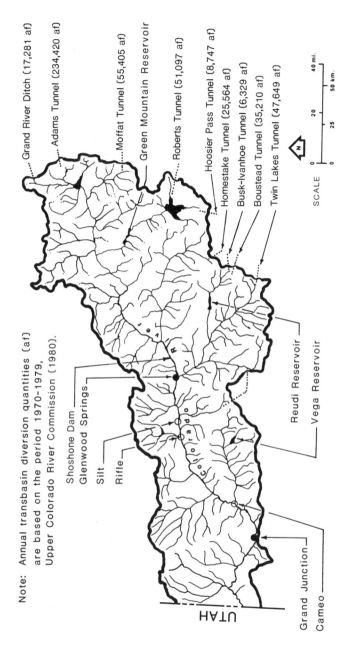

Fig. 6. Map of Water Division 5 and transbasin diversion points

common in water disputes, the precise operating policy for **release** of replacement storage waters has yet to be formulated.[48] Water rights for Denver's reservoirs on the Blue River (i.e., Dillon, Two Forks, and Marston reservoirs), are laid out in the "Blue River Decree." These two documents provide the basic operating guidelines for regulating streamflow in the headwaters of the Colorado River main stem. Dispute still exists, however, over whether *any* downstream call on the river can trigger release of replacement waters or whether the call must be made at the Dotsero gauge for the Shoshone Power Plant and Dam (see figure 6).

More recently, a municipal subdistrict of the Northern Colorado Water Conservation District entered into an agreement with its Western Slope adversary, the Colorado River Water Conservation District, on the Windy Gap Project, a major diversion project for Front Range municipal use. The precedents established in Senate Document 80 and the Blue River Decree along with additional provisions for the maintenance of in-stream flows and satisfaction of environmental permits were incorporated into the agreement.[49]

Transbasin diversion into the Arkansas River Basin was authorized in the Bureau of Reclamation's Fryingpan Arkansas Project in 1962, subject to operating principles adopted by the United States and State of Colorado in 1959.[50] In this case Reudi Reservoir was constructed on the Fryingpan River to provide replacement storage, in-stream flow regulation, and a marketable source of water for various uses. Until quite recently, none of the 40,000 acre-feet allocated to these uses had been sold. Recent contracts with the Colony Project led to competition among local and regional entities for the authority to market Reudi water. The town of Aspen sought control of the reservoir, presumably favoring the recreational and ecological components of multiple use, while the Colorado River Water Conservation District sought control of the reservoir presumably as a source of funds to construct additional Western Slope storage facilities. Thus far the Bureau has retained control of the reservoir and has let the first three contracts for its water: to the West Divide Water Conservancy District; the Basalt Water Conservancy District; and to the Colony Project with

---

48. Bureau of Reclamation, "Colorado-Big Thompson Project; Proposed Green Mountain Reservoir Operating Policy," *Federal Register* 46:58 (1981): 18801-2.

49. "Agreement concerning the Windy Gap Project and the Azure Reservoir and Power Project," reprinted in U.S. Department of the Interior, Bureau of Reclamation, *Final Environmental Statement: Colorado-Big Thompson Windy Gap Projects, Colorado* (Washington, D.C.: Government Printing Office, 1981).

50. Frying Pan-Arkansas Project Act of 1962, P. L. 87-590, 76 Stat. 389; and U.S. House of Representatives, *Operating Principles, Frying Pan-Arkansas Project,* Doc. 130, 87th Cong., 1st Sess. (1961).

Battlement Mesa Inc., a new town planned by Colony.[51] Given the favorable terms under which these contracts were let, further purchases of Reudi water for municipal and industrial use seem likely. Contracts for Green Mountain Reservoir water, also available, are much less likely, however, due to the imbroglio of legal action pending on the allocation of that water.

On the main stem itself the major senior decrees belong to the Grand Valley Irrigation Company, the Grand Valley Project, and the Shoshone Power Plant and Dam (figure 6). When any of these rights is shorted, the Division Engineer puts a "call" on the river, meaning that Green Mountain Reservoir or in some cases the Williams Fork Reservoir must release water. In the past, if the Shoshone Power Plant decrees were fulfilled, then downstream senior decrees were also fulfilled. Construction of major diversions and storage reservoirs at Parachute and Roan Creeks and reservoirs along tributaries such as Elk and Canyon Creeks could overturn this simple operating guideline and lead to controversy in years when all replacement storage from Green Mountain has been released or when main stem decrees other than the Shoshone Power Plant are shorted.

Local Water Control

Whereas State agencies have historically fulfilled administrative functions at the local level, a host of public and private entities have actually developed and managed the water. At the broadest level, the Colorado River Water Conservation District conducts legal action, engineering studies, project development, and policy analysis for that area of the Colorado River Basin within northwestern Colorado. The district was established in 1965 and is funded by special property assessments.[52] It has actively promoted Western Slope water development and the protection of Western Slope water interests against downstream appropriators and transbasin diverters.

Financing for the earliest appropriations came from irrigation mutuals and speculative ditch companies.[53] Moses suggests that problems of capital formation and risk on larger projects generated a sequence of institutional innovations leading from unincorporated ditch companies to irrigation districts and, subsequently, to water conservancy districts.[54] Briefly, irrigation districts are

---

51. Draft contracts dated 9/28/81 to the West Divide Water Conservancy District and the Basalt Water Conservancy District; draft contract dated 10/23/81 for the Colony Project and Battlement Mesa, Inc.

52. C.R.S. 1973, 37-46-101, et seq.

53. C.R.S. 1973, 7-42-101 et seq.

quasi-public corporations with power to levy assessments on irrigated land for the construction and maintenance of irrigation facilities in the district.[55] Water conservancy districts are quasi-municipal corporations, organized to finance and operate larger scale multiple purpose projects and empowered to levy assessments on all lands in the district. Five water conservancy districts (WCD's) lie within the study area--Basalt WCD, West Divide WCD, Silt WCD, Collbran WCD, and the Battlement Mesa WCD.[56] Water conservancy districts, irrigation districts, and water users associations, such as the Grand Valley Water Users Association, may enter into contracts with the Bureau of Reclamation on Federal water projects.[57] The historical importance, powers, and limitations of various organizational forms receive more complete treatment in chapter eight. Here we have sought only to carry the discussion of water management and water control down to the local level.

Introduction to the Study Area

The drainage basin of the Colorado River main stem from Glenwood Springs downstream to the Utah border (excluding the Gunnison River Basin), comprises an area of approximately 3,825 square miles (figure 7). This downstream segment of Water Division 5 has experienced extensive pressure on agricultural water use as a result of oil shale development and urban growth. To be sure energy firms have also acquired water rights on the Roaring Fork and Gunnison rivers; transbasin diversions have been proposed from the neighboring Yampa and White rivers; and storage rights options have been acquired for reservoirs outside the study area. The greatest changes in traditional water use patterns, however, are expected to occur along this corridor between Glenwood Springs and Utah.

The Colorado River follows a narrowly winding and constricted path for thirteen miles through Glenwood Canyon east of the study area. Shoshone Dam, discussed as one of the more important water rights on the main stem, impounds the flows of the Colorado halfway through Glenwood Canyon and then releases them through the penstocks of its power plant several miles downstream. Another six miles downstream the canyon walls recede, and the Roaring Fork River discharges into the Colorado.

---

54. Raphael Moses, "Irrigation Corporations," *Rocky Mountain Law Review* 32 (1960): 527-533.

55. C.R.S. 1973, 37-41-101 et seq., "Irrigation District Law of 1905;" 37-42-101 et seq., "Irrigation District Law of 1921;" 37-43-101 et seq., "Irrigation District Law of 1905 and 1921."

56. C.R.S. 1973, 37-45-101 et seq.

57. C.R.S. 1973, 7-42-101 et seq., "Water User's Associations;" and C.R.S. 37-45-101 et seq., "Water Conservancy Districts."

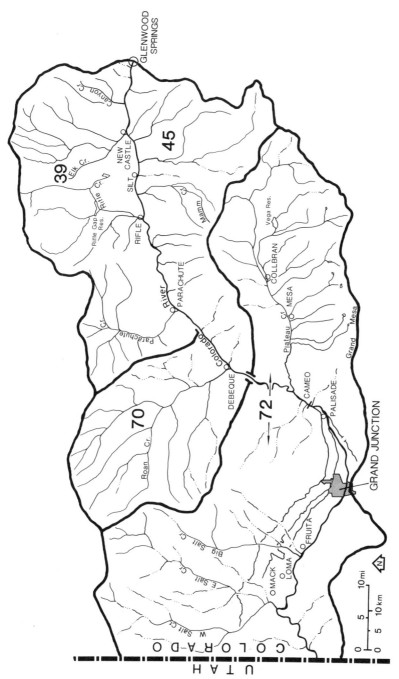

Fig. 7. Map of the study area and substate water districts

The confluence of the Colorado and Roaring Fork rivers at Glenwood Springs marks the eastern boundary of the study area. Between Glenwood Springs and Grand Junction the river drops in elevation from 5,823 to 4,602 feet. A precipitation gradient follows this trend in elevation quite closely, ranging from an annual average of 16.46 inches at Glenwood Springs to 8.41 inches at Grand Junction.[58] As a result, vegetation patterns also trend along a gradient from mesic associations in the east to more xeric associations in the west. Physiography of this region has been established by the simple folding and faulting of three major plateaus: the White River, Roan, and Grand Mesa. Fluvial action by the Colorado and its tributaries has carved narrow canyons from these plateaus and has built fertile floodplains which vary greatly in width but which rarely exceed four miles.

Irrigation activity is concentrated on these floodplains as well as on lower mesas and river terraces. Most farm products are related to livestock production--both sheep and cattle. Principal irrigated crops are hay, alfalfa, and other forage crops, although smaller areas have produced orchard fruits--principally apples, pears, cherries, and peaches--since the turn of the century.

Irrigation agriculture has had a relatively recent history in this region. Ute Indian groups did not irrigate; they hunted throughout west-central Colorado, prior to the "Ute Removal" in 1881, coexisting with early trappers and drawn to the putatively healing powers of mineral waters at Glenwood Springs. Settlement in the region began with trapping and proceeded through a sequence of mineral booms and ranching ventures. Early agriculture provided support for these endeavors and naturally followed their vicissitudes as well. With the expansion of urban settlements in the West and the extension of livestock and fruit production into larger regional markets, irrigated agriculture began to stabilize. Even today, however, irrigated crops continue to serve primarily local demand for forage and produce.

Agricultural production in the region is concentrated along the Colorado River floodplain and irrigable portions of tributaries and side-canyons. A string of small towns lie along the main stem between Glenwood Springs and Grand Junction--situated at the confluence of irrigated tributaries with the Colorado. These towns vary in size with the scale of irrigation activity, and they provide local road and rail connections for commerce along the major

---

58. U.S. Weather Bureau, *Climatological Data,* annual summaries 1902-1978.

east-west regional corridor. The larger towns also serve local mining operations, government offices, and recreational activities. Glenwood Springs in particular serves the recreation and tourism industry, whereas Grand Junction functions as a more diversified regional center for mining and agriculture.

Until recently the most dramatic mineral developments had occured just outside the study area. Mineral development within the study area began with the first oil shale boom in the 1920's, followed by uranium mining near Grand Junction in the 1950's, coal mines in DeBeque Canyon, and most recently commercial oil shale development centered on the Parachute and Roan Creek drainages.[59]

Acquisition of water rights for energy uses began in the late 1940's as firms sought both conditional water rights and senior irrigation rights to satisfy anticipated production requirements. This intensified in the 1960's and 70's and then tapered off as energy water demand estimates were adjusted downwards and the commercial water demands of most projects appeared to be more than adequately met. Few firms acknowledge a current active pursuit of senior irrigation water rights. Most claim this had never been a central part of their acquisition strategy, though in a number of cases the record suggests otherwise.[60] The present disinterest in senior rights has several underlying causes: first, "drying up agricultural lands" has become a very sensitive political issue in the region and the State; and second, early water transfer attempts gave harsh lessons of how the "no injury" and "historic beneficial use" criteria for water rights changes can drastically reduce the volume and value of senior rights.

The impacts of energy development on irrigation agriculture, however, are far from simple. Sale of land and water to energy firms is often accompanied by lease-back arrangements with farmers. Sometimes this enables consolidation of farms or upgrading of water control devices, but in other cases it results in inefficient fragmentation, exacerbated conflict, and neglect of local water management responsibilities. Soil conservationists are undecided on the impacts that such arrangements have had on investment in conservation practices by the new landlords and their tenants.[61] The

---

[59]. More detailed historical treatment of mineral development on the Western Slope may be found in Duane VandenBusche and Duane A. Smith, *A Land Alone: Colorado's Western Slope* (Boulder, CO: Pruett Press, 1981); and Steven F. Mehls "The Valley of Opportunity: A History of Western Colorado," unpublished draft (Denver: Bureau of Land Management, 1980).

[60]. Witness Union Oil's transfer of senior water rights from the Roaring Fork drainage over 40 miles downstream to Parachute Creek.

[61]. Interviews with field conservationists in Glenwood Springs and Grand Junction Soil Conservation Service offices, 1981.

Colorado Department of Agriculture states that a "cost-price squeeze" has driven many farmers and ranchers out of business, independent of the energy boom.[62] In fact, it has been suggested that escalating land values have enabled some farmers to increase their credit and subdivide smaller portions of their farms in order to continue farming. While it may be difficult to separate out the supportive and erosive effects of energy development on agriculture, it is quite clear that agriculture is both shrinking and changing within the study area.

With this broad picture of the study area as a regional corridor which tenuously supports irrigation farming along the spine of the Colorado and its numerous minor tributaries--an area beset with the problem of providing water for emerging energy development--we may proceed to a finer level of description. Because much of this discussion will focus on differences in water use patterns and problems among the four administrative water districts which lie in the study area, a brief presentation of each district is provided below (figure 7).

Water District 39

As the combined flows of the Colorado and Roaring Fork rivers proceed westward from Glenwood Springs, they receive the relatively minor flows of numerous smaller tributaries from both the north and south. The River serves as an administrative boundary in this upstream portion of the study area. Drainages lying north of the Colorado from Oasis Creek near Glenwood Springs downstream to Parachute Creek are administered within district 39, and watersheds south of the River comprise district 45.

The smaller tributaries in district 39 (such as Oasis, Mitchell, Canyon, and Government) support small-scale relatively unsophisticated hay crop irrigation and horse pasturage. More intensive hay and alfalfa irrigation occurs along the Colorado River from Canyon Creek to Parachute Creek, in several cases employing sideroll sprinkler systems and gated pipe. These water systems are either gravity-fed by tributary diversions or pumped from the Colorado. Construction of tributary storage reservoirs behind the Grand Hogback has improved irrigation agriculture in the larger tributary floodplains of Elk and Rifle Creeks. Although these areas also employ unlined ditches and produce forage crops, the scale of farming, length of the irrigation season, and quality of farm products exceed those of the smaller drainages. Completion of the

---

62. Colorado Department of Agriculture, *Agricultural Land Conversion in Colorado,* 2 vols. (Denver, 1980).

Bureau of Reclamation's Silt Project in 1967, for example, greatly improved water supplies for an irrigable area of 6,591 acres along the Colorado River near Silt and north of the Grand Hogback near the Harvey Gap Reservoir.[63]

Water rights administration in district 39 is concentrated first, on creeks which run dry each year such as West Rifle and Government creeks; and second, on areas of West Elk Creek where personality conflicts and allegations of over-watering and water misuse in the Silt Project have generated disputes.[64] Rapid suburbanization of irrigated farmland has occured near the town of Silt (1980 census population--923) which serves as a retail service and residential center near the Silt Project, and also near the larger town of Rifle (1980 census population--3,215) which has become the fastest growing residential and commercial center between Glenwood Springs and Grand Junction.

Continuing westward, vegetation changes give clear evidence of the precipitation gradient from mesic upstream locations near Glenwood Springs to more xeric conditions near Parachute Creek. Relatively dense stands of scrub oak *(Quercus gambellii)* on hillslopes, pinyon-juniper communities *(Pinus edulis, Junipus osteosperma)* on rock outcrops, and cottonwood-chokeberry *(Populus angustifolia, Prunus virginiana)* along stream bottoms gradually give way to bluebunch wheatgrass *(Agropyron spicatum),* Indian ricegrass *(Oryzopsis hymenoides),* and big sage *(Artemisia tridentata)* communities in downstream locations.

Parachute Creek, the most westward of the district 39 tributaries, has historically supported a narrow band of irrigated haylands along its floodplain. The creek provides passage between grazing lands on Roan Plateau and the rural service center of Parachute (1980 census population--338) at its mouth. Dramatic changes along Parachute Creek seem imminent: construction has begun on the first two commercial-scale oil shale mines; a four lane road now runs up Parachute Creek; and roads have been laid for a new town of 25,000 people across the river on Battlement Mesa. In 1981 a number of leased back farms and water rights were called in by energy firms, giving actual substance to the changes forecasted to accompany oil shale development.

---

63. U.S. Bureau of Reclamation, "Silt Project," in *Project Data* (1980).

64. Interviews with the Division 5 Engineer and Water Commissioners.

Water District 45

On the south side of the Colorado River, opposite district 39, lies a set of roughly parallel drainages running from South Canyon Creek, just west of Glenwood Springs downstream to Spring Creek at the entry to DeBeque Canyon. Several small transbasin diversions have been made from watersheds south of the study area into district 45.

As in district 39, conflicts are most intense on irregular streams such as Beaver, Cache, and Porcupine Creeks and in basins where personality clashes have developed.[65] Water supplies are most reliable in larger drainages such as Divide Creek and Mamm Creek. Denser vegetation in the northern-oriented drainages of district 45 is offset by shorter stream lengths, but otherwise the two water districts have similar elevations and precipitation patterns.

The most important distinctions between irrigation practice in districts 39 and 45 concern water storage and types of irrigated land. The lack of storage in district 45 is often cited as the cause for lower irrigation rates and crop yields. The observation that hay crops "brown out" in district 45 and allow fewer cuttings may be explained in part by the absence of storage, but must also consider the dispersion of irrigation activity over broad upland mesas of district 45 where conveyance distances are longer and losses correspondingly greater. Return flows from these mesas are likely to be inconsequential or very slow in comparison with the valley bottom irrigation characteristic of district 39. Plans to provide storage for district 45 through the Bureau of Reclamation's West Divide project have been pending for roughly fifty years--impeded by reports of poor feasibility, environmental protest over proposed storage reservoirs, and inadequate local support to secure financing.[66]

No towns lie within district 45, but subdivisions have spread south from Rifle and west from Glenwood Springs. Local services, and highway and rail connections all must be obtained across the river in New Castle, Silt, Rifle, and Parachute. Many farms and water rights, while relatively remote from the nexus of oil shale development, have been sold to developers, speculators, and energy firms. Construction of the Battlement Mesa new town across from Parachute Creek constitutes the first major settlement in the district.

---

65. Interviews with the Division 5 Engineer and Water Commissioners (1981).

66. U.S. Bureau of Reclamation, "West Divide Project" in *Project Data* (1980); and idem, *Potential Modification in Eight Proposed Western Colorado Projects for Future Energy Development* (1980): 30-37.

Water District 70

The Roan Creek basin lies north of the Colorado River adjacent to Parachute Creek. In physiography and oil shale resources the two basins have many similarities, but the large size of Roan Creek led to its designation as a separate water district. Owned in large part by energy firms, irrigated farms continue to produce forage crops for winter feed on leased-back lands. Roan Creek provides a much broader valley bottom for irrigation than Parachute Creek, but many of its tributaries such as Conn, Kimball, and Brush Creeks irrigate only narrow ribbons of floodplain constrained by steep talus slopes and intermittent streamflows.

Water conflicts are said to be few at present in the Roan Creek drainage, in part due to consolidation of farms and in part due to the ownership by energy firms which are said to be conscientious in installing measuring devices and in mitigating disputes.[67] In spite of this, water management remains crude relative to other districts, reportedly due to the difficulty of controlling washouts and shaley debris pile-ups in water control structures. Few formal water organizations exist in the district, and simple ditch agreements still serve to fulfill annual maintenance and repair needs. Frequently all this entails is hiring a local backhoe operator to clean debris out of a ditch. Only the Bluestone Water Users Association, located along the Colorado River, operates with by-laws, stockholders' proceedings and formal assessments. This association has recently joined with the Battlement Mesa Water Conservancy District, the West Divide Water Conservancy District, and the small town of DeBeque at the mouth of Roan Creek to seek new project water. The town of DeBeque resembles Parachute in many ways but does not have as severe a land constraint, nor so imminent a prospect of rapid growth related to oil shale development--one formally announced project of 50,000 bbl/day by Chevron was scheduled for construction on Clear Creek.

Water District 72

At the western boundary of district 70 the Colorado River enters the narrow Debeque Canyon. While not as dramatic as Glenwood Canyon, perhaps, Debeque Canyon poses an equally effective physiographic constraint on irrigation. District 72 thus begins at the entry to Debeque Canyon and spans both sides of the Colorado River westward to the Utah border. District 72 comprises two very different water use areas: first, the Plateau Creek drainage, and

---

67. Interview with District 70 water commissioner.

second, the Grand Valley irrigation area along the Colorado River.

Plateau Creek rises from an elevation of 4,800 feet at its mouth near Cameo to over 10,500 feet in its headwaters. Like so many other drainages in the study area, the Plateau Creek basin supports mainly forage production and irrigated pasture for cattle that graze on Bureau of Land Management and Forest Service lands on the Grand Mesa plateau. Collbran, the principal town in the drainage, functions as a rural and recreational service center as do a handful of smaller settlements. Upstream from Collbran lies Vega Reservoir, part of the Bureau of Reclamation's Collbran project completed in 1960, which provides full or supplemental service for 22,210 irrigated acres, as well as flood control, recreational, and power generation benefits.

Water use along the Colorado River main stem in the Grand Valley area stands in marked contrast with that of Plateau Creek and other districts. The Grand Valley area encompasses a band of irrigated land roughly four miles wide by 75 miles long that extends from the mouth of Debeque Canyon near the town of Palisade downstream to the town of Mack. Although small in comparison with irrigated areas in eastern Colorado, the Grand Valley area contains the oldest, largest, and most sophisticated irrigation activities in western Colorado. Orchard cultivation, which once occured as far east as Glenwood Springs, is now concentrated around the town of Palisade where air drainage through DeBeque Canyon provides protection against frost damage. Relatively diversified cultivation of forage crops, corn, barley, wheat, and soybeans takes place in central and western Grand Valley.

The common statement that irrigators in Grand Valley hold the first two decrees on the Colorado River overgeneralizes a fascinating record of water development. Irrigation was practiced by numerous private ditch compaines in the early 1880's. Consolidation of these led to construction of the privately financed and cooperatively organized Grand Valley Canal--the first major decree on the river for 520.8 cfs during the growing season. Early requests in the 1890's for a State-financed highline canal were followed by an unsuccessful Federal attempt to sponsor a project only one month after passage of the Reclamation Act of 1902.[68] Conflicts over title, institutional organization, and repayment were resolved in 1911. Water was finally delivered to the Grand Valley Project through the Government Highline Canal in 1915, and in 1930

---

68. U.S. Department of the Interior, *Federal Reclamation and Related Laws Annotated,* vol. 1 (Washington, D.C.: Government Printing Office, 1972), p. 52.

the financially troubled Orchard Mesa Irrigation District was added to the Project. Water rights for the Federal project were consolidated, thus establishing the second right on the River of 1,625 cfs during the irrigation season and 800 cfs for power production in winter months. Today there are four major agricultural water districts, three of which receive water from the Bureau of Reclamation's Grand Valley Project.

Drainage problems surfaced by 1911 as a result of canal seepage and overirrigation, leading to formation of the Grand Valley Drainage Association. The Bureau of Reclamation financed a major drainage ditch construction effort from 1919 to 1921. More recently, the 1974 mandate to reduce river salinity has led to the Bureau's involvement in a canal lining program for the Grand Valley Salinity Control Project.

Grand Junction, which lies at the confluence of the Gunnison and Colorado Rivers, has become the largest commercial and residential center in western Colorado. Rapid municipal growth has extended to the surrounding municipalities of Palisade, Clifton, and Fruita. These municipalities receive water from reservoir and direct flow sources on Grand Mesa, Kannah Creek, and Plateau Creek. Domestic water users in unincorporated areas are served by the Ute Water District, a special purpose rural domestic water district holding water rights on Plateau Creek and Jerry Creek. Exurbanization and sprawl have led to overextension of the Ute system's capacity. Planners argue that extension of the Ute system has in several cases produced inefficient growth patterns. The Ute system originated to serve rural domestic uses and has not been able to supply the fire protection and pressure needs of its expanded urbanizing service area. Coordination of rural domestic water supply and county planning objectives is seen as an important need by public officials.

Municipal growth has also generated water management problems in irrigated areas, particularly in Grand Valley Project lands that lie under the Government Highline Canal and the Orchard Mesa Canals. Problems revolve around shifts in the timing of demand (peak demands are more severe), disrespect of allocation rules within the service area, and the failure of residential water users to assume basic maintenance responsibilities on ditches and structures--issues that water managers and farmers attribute to "non-western" attitudes toward water rights and water availability. In the case of Orchard Mesa this sometimes results in upstream-downstream conflicts along canals and laterals that necessitate considerably more labor by water managers and ditch riders. Some subdivisions and farms are said to hire out their own ditch riders to protect their rights within the system (note--once the water leaves a public stream, its distribution becomes a private problem for the water users). Suburban water users pay larger amounts per unit of water but have no representation in the water districts. By-laws of agricultural water organizations have historically limited voting rights to land holders of one acre or more. On Federal project lands the problem is most severe because water remains appurtenant to specific tracts of land at a constant duty of water, regardless of changes in land use, whereas in private organizations, stock may be transferred among users within the service area. Clearly, the transition from single purpose agricultural water management to combined irrigation, lawn watering, and general outdoor use systems represents a major water management problem in the district.

Irrigators sometimes hesitate to consider conservation options for fear of challenges to their water rights over the issues of overappropriation, over-irrigation, or other cases in which water may not have been fully applied to beneficial use. These fears are responsible in part for the local practice of keeping the canals brim full, the headgates wide open, and the fields flooded with the "most senior rights on the Colorado River."[69] The Colorado River, laden with saline return flows from the Grand Valley area and the Gunnison River, then continues in a southwesterly direction through the narrow valleys and canyonlands of southeastern Utah until it is received with great anticipation by the Lower Basin.

Summary

Water use studies often begin with a ritual recitation of the "law of the Colorado River"--its compacts, lawsuits, and administration; followed by an outline of State water law and administrative organization. This then sets the stage, as well as the assumptions, uncertainities, and constraints for water forecasting models. In expanding on this tradition, three additional purposes have been served. First, while no means exhaustive, the narrative has sought to convey a broad sense of the expression "Colorado River System," first mentioned in the Colorado River Compact of 1922, so that local activities and plans are always

---

69. Telephone interview with a representative of the Palisade Irrigation District. These fears were reinforced by the cancellation of conditional water rights held by the Orchard Mesa Irrigation District *(Orchard Mesa Irrigation District v. City and County of Denver,* 182 Colo. 59, P.2d 25 (1973)) in 1973 and more recently of those held by the Colorado River Water Conservation District, see "Colorado Supreme Court Upholds Decision on Western Slope Water," *Glenwood Post* (Colorado), 8 February 1982; and *Colorado River Water Conservation District v. City and County of Denver,* ___ Colo. ___, 640 P.2d 1139 (1982).

viewed in a global context. Second, we have emphasized that the hierarchical elements of water control do not merely establish constraints on local action, they alter it by invoking strategic behavior among participants in the allocation process. Finally, the scoping process has conveniently illustrated the local complexities underlying generalizations and summary statistics at larger scales of analysis.

The scarcity theme came out most forcefully in studies of the entire basin. Assessments of Upper Basin water supplies, on the other hand, explore various paths for development of a somewhat embarassing water surplus. At an even finer level of observation, that is the sub-basin and tributary basin scale with which this research is principally concerned, however, a complex mosaic of scarcity and surplus emerges. In the more intensive analysis that follows, it becomes increasingly clear that along tributaries and even larger canals scarcity and surplus are concepts shaped by institutional, locational, physiographic, technological, and cultural circumstances, both locally and in larger arenas.

CHAPTER IV

PATTERNS OF WATER RIGHTS APPROPRIATION

In this chapter the record of water rights appropriation is evaluated as a means of interpreting patterns of water development in the study area. In an early assessment of western water law, Elwood Mead credits Colorado as, "having been the first State to enact a code of laws for the public administration of streams," but he faults Colorado's codification of the appropriation doctrine for its inadequate consideration of either the public interest or the needs of future irrigators--problems he interprets as arising from excessive water rights appropriation and inadequate administrative procedures.[1] As Mead notes, water rights patterns do not accurately portray actual water demand, but rather represent a cumulative legal record of water development. The pattern of absolute flow rights may vary significantly from actual patterns of water diversion. Similarly, the pattern of conditional decrees represents an array of water development alternatives and not an exact projection of future water demand.

Why consider water rights at all then? Although water rights analysis cannot by itself forecast water availability, it can offer valuable information on the record, pattern, and types of water development within a region. Moreover, the relationships among water rights in a hydrologic system have a direct bearing on the ability to alter those water rights or to transfer them to other uses. As a result, the geographical patterning of water rights at any point in time often affects subsequent water acquisition strategies.

In chapter three major water control points along the Upper Colorado River main stem were noted: Green Mountain Reservoir, Shoshone Power Plant and Dam, Reudi Reservoir, the Cameo Diversion

---

[1]. Elwood Mead, *Irrigation Institutions* (New York: Macmillan, 1903), p. 143.

Dam, and the Grand Valley Diversion Dam (figure 6). The more exhaustive investigation of agricultural and industrial water rights which follows describes their distribution in 52 stream basins which lie within four State water districts (figure 8 and table 1). Discussion centers first on the uses and limitations of water rights tabulations, the principal source of information on water rights. Description of water rights patterns then turns to the timing of appropriation; and finally, to the association between the intensity of appropriation and key physical, economic, and institutional variables. This detailed description of water rights patterns provides a foundation for discussion of water rights filings for energy development.

TABLE 1

STREAM BASINS INCLUDED IN THE ANALYSIS

A. District 39

1. Oasis Creek
2. Mitchell Creek
3. Canyon Creek
4. Elk Creek
5. Rifle Creek
6. Government Creek
7. Parachute Creek
8. Colorado River (north side)

B. District 45

9. Garfield Creek
10. Divide Creek
11. Dry Hollow
12. Mamm Creek
13. Dry Creek #1
14. Cache Creek
15. Porcupine Creek
16. Beaver Creek
17. Battlement Creek
18. Dry Creek #2
19. Spring Creek (45)
20. Wallace Creek
21. Colorado River (south side)

C. District 70

22. Clear Creek
23. Brush Creek (70)
24. Carr Creek
25. Roan Creek
26. Kimball Creek (70)
27. Dry Fork
28. Conn Creek
29. Colorado River

D. District 72

30. Big Salt Creek
31. East Salt Creek
32. West Salt Creek
33. Colorado River
34. Plateau Creek
35. Rapid Creek
36. Mesa Creek
37. Coon Creek
38. Spring Creek (72)
39. Bull Creek
40. Cottonwood Creek
41. Buzzard Creek
42. Kimball Creek (72)
43. Big Creek
44. Grove Creek
45. Salt Creek
46. Hawxhurst Creek
47. Brush Creek (72)
48. Leon Creek
49. Owens Creek
50. Collier Creek
51. Cheney Creek
52. Yule Creek

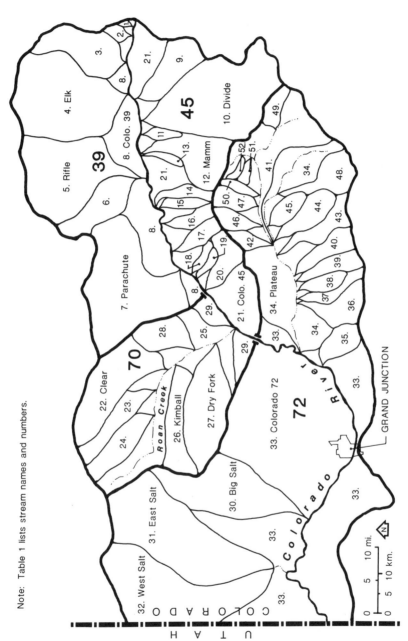

Fig. 8. Map of stream drainages

Water Rights Tabulations

County records of water rights appropriation were not required until 1881, and State records were only initiated after 1887.[2] The principal use of these records has lain not in comprehensive management of the public's resources, but in the resolution of private disputes by water commissioners in the field and water lawyers in the courts. River basin planning studies conducted prior to 1969 had to laboriously compile these decree records, a good number of which are exaggerated or "paper" rights in Mead's sense.

Systematic and periodic tabulation of water rights decrees was initiated to mitigate these difficulties after 1969 with the Water Right Determination and Administration Act.[3] This chapter relies in large part on the 1978 Division Engineer's Water Rights Tabulation for streams in Districts 39, 45, 70, and 72. The following information is recorded on each absolute and conditional water right in the tabulation:

1. *Name of Structure*
2. *Type of Structure* (ditch, reservoir, pipeline, well, spring)
3. *Name of Source* (stream name, well number)
4. *Location* (district, township, range, section)
5. *Type of Use* (irrigation, municipal, commercial, industrial, recreation, stockwatering, etc.)
6. *Amount Decreed* (flow rights are specified in cubic feet per second while storage rights are measured in acre-feet)
7. *Type of Adjudication* (original, supplemental, transfer, abandonment, alternate point of division)
8. *Adjudication Date*
9. *Appropriation Date*
10. *Basin Rank* (by Division)
11. *Engineer's Notes*.

Although an advance over previous compilations, the current tabulation for Colorado Water Division 5 has a number of important limitations that affect its accuracy and coverage. Many water court decrees and case actions of the past decade, for example, have yet to be tabulated. Errors and omissions in the location of diversion structures, adjudication dates, and the type of decree are commonplace.[4] The loose administrative record of Division 5 stands

---

2. Del Breese Kincaid, *Irrigation Law of Colorado* (Denver: W.H. Courtright Publishing Company, 1912).

3. C.R.S. 1973, 37-92-101 et seq. as amended in 1971, and 1975. John H. Carlson, "Water Tabulations and Abandonment" in *Water Law 1978*: *Proceedings* (Denver: Colorado Bar Association, 1978).

4. Personal communication with the Division Engineer and the Assistant Engineer.

in marked contrast with the activities of other divisions. In
Division 6, for example, structures and conveyances are mapped,
irrigation requirements estimated, and land uses surveyed to
facilitate a more efficient resolution of water rights conflicts in
the field and in the courts.[5]

Abandonment lists have not been prepared for any of the
Colorado River drainage divisions, as required by the Water Right
Determination and Administration Act. Protection of the State's
compact apportionment has superseded all such issues of
administrative efficiency and conservation.

> "Abandonment of a water right" means the termination of a water right
> in whole or in part as a result of the *intent* of the owner thereof to
> discontinue *permanently* the use of all or part of the available water
> thereunder. (C.R.S. 1973, 37-92-103, emphasis added).
>
> . . . failure for a period of ten years or more to apply to a
> beneficial use the water available under a water right when needed by
> the person entitled to use same shall create a rebuttable presumption of
> abandonment . . . (C.R.S. 1973, 37-92-402(11), as amended).

Proving an intent to permanently abandon a water right becomes an
extremely difficult task, and in one case, non-use of a water right
for 26 years was not held to conclusively prove its abandonment.[6]
Conditional decrees, also listed in the tabulation, may be cancelled
for failure to exercise "due diligence" on a quadrennial basis
(C.R.S. 1973, 37-92-103).[7]

More significantly, the 1969 Water Right Determination and
Adminstration Act does not require the type of information necessary
to fully describe the legal pattern of water supply and demand in
any given area. The State administers only the withdrawal of water
from public streams and aquifers. It does not record the location
of use, type of crop or industry, number of municipal users, acreage
irrigated, means of conveyance, or distribution of water among users
after the water is withdrawn from a public water source. The
massive administrative system that would be necessary to collect
such information and an ideological opposition to the State "poking
its nose" into the use of private water rights have historically
limited the State's role in water management.

---

5. Kent Holt, "Division No. 6 - Water Budget Program," Division 6
Engineer's Annual Report (Steamboat Springs, CO, 1980).

6. *Orr v. Denver,* 572 P.2d 805 (1977).

7. *Denver v. Northern Colorado Water Conservancy District,* 130 Colo. 375,
276 P.2d 992 (1954). The diligence concept has shifted from an original emphasis
on physical construction to a broader range of activities including technical
assessments and procedural compliance. More rigorous diligence requirements may
be forthcoming as competition for water steps up in the Upper Colorado River
Basin. See *DeBeque v. Enewold,* 606 P.2d 48 (1980); and *Simineo v. Kelling,* 607
P.2d 1289 (1980); and most recently *Colorado River Water Conservation District v.
City and County of Denver,* 640 P.2d 1139 (1982).

These limitations do not preclude investigation of water rights patterns on a regional level or the use of the water rights tabulation; but they do necessitate caution in such investigations. The most extensive application of water rights analysis has been undertaken by a consortium of energy firms, water districts, and major utility companies which produced the Colorado River Simulation Model (CORSIM). All tabulated decrees greater than 3.5 cfs were field checked and entered into a simulation model designed to assess the yield of water rights holdings at any location in the stream network and under varying hydrologic and water demand conditions.[8] Several engineers and water lawyers have pointed out, however, that although such studies have value for regional water forecasting, the assessment of individual water rights requires much more intensive investigation of the right, its locational context, and the historical record of its use.[9] The present description of water rights patterns lies in between the basin-wide scale of CORSIM and the intensive investigation of individual rights.

Water Rights Patterns

Water rights are unevenly distributed in time and space--a fact which takes on special importance in evaluating proposals to meet increasing or shifting water demands. The size and priority of water rights relative to local physiographic and economic conditions directly affect the legal availability and relative value of different water supplies. At present, regional description of water rights patterns is necessarily crude for the reasons discussed above. Nevertheless, a summary description of the types of water rights, timing of appropriation, and intensity of water development in the study area, however crude, constitutes an essential prelude to the consideration of water acquisition strategies.

At first glance it becomes apparent that most rights in the study area are decreed for irrigation. In contrast with eastern Colorado, relatively small amounts of water have been appropriated for municipal, domestic, stockwater, and recreational uses. Exceptions include the water rights for Rifle, New Castle, Silt, Parachute, and other small municipalities. Glenwood Springs water supplies come from Grizzly and No-Name Creeks in Glenwood Canyon, while Grand Junction water comes from the Gunnison river and Kannah

---

8. David E. Fleming Co., "The CORSIM Project," paper presented to the annual meeting of the American Society of Civil Engineers, Denver, CO., 1975. The proprietary data in this model are unfortunately not available for public distribution.

9. See for example, Wright Water Engineers, "Water Rights and Water Resources Analysis for Battlement Mesa, Inc., and Battlement Mesa Farm Lands," Glenwood Springs, CO., 1977. (Typewritten.)

Creek, a tributary of the Gunnison. Only the Ute Water District, a rural domestic water supplier, obtains water of any significant quantity from the study area (Plateau Creek drainage).

Unlike other water rights, irrigation decrees extend only through the growing season. At present no restriction or "duty of water" limits the amount of irrigation water that may be appropriated for an acre of land, except on U.S. Bureau of Reclamation project lands. The ability to transfer water from irrigation to urban or industrial use is limited, however, by both the seasonal length and crop requirements of standard irrigation practice--an issue taken up in detail in chapters six and seven.

The distribution of appropriations through time follows a common path, beginning in the 1880's in most basins (see figure 9). Absolute water rights represent a traditional pattern of appropriation, dominant from 1880 to 1920. Smaller basins such as Yule Creek were fully appropriated in the second decade of this century, and only the larger streams continue to be appropriated. Exceptions include Leon Creek whose appropriation coincides with the development of the Collbran Project after 1950. Nevertheless, in most streams fifty to ninety percent of absolute flow rights were appropriated by 1920. Conditional rights and storage rights are relatively recent phenomena, reflecting shifts in water demand that have occurred over the past thirty years.

Water rights in this region fall into four major groups based on the status of the decree (absolute or conditional) and the nature of the water source (direct flow diversion from wells and streams or reservoir storage water).

1. *Absolute direct flow decrees (AFLOW)*--a water right perfected by actual use and measured as a rate of flow in cubic feet per second (cfs)
2. *Conditional direct flow decrees (CFLOW)*--a water right granted on legal demonstration of intent to put water to use, contingent upon demonstration of due diligence, and measured as a rate of flow (cfs)
3. *Absolute storage decrees (ASTOR)*--a perfected right to store water in a reservoir, measured as a volume in acre-feet (af).
4. *Conditional storage decrees (CSTOR)*--a right granted on demonstration of intent and due diligence to store water at some future date

The Western Slope is distinguished from the Front Range of the Rockies by its predominance of direct flow rights and relative absence of storage rights--a difference interpreted again as the

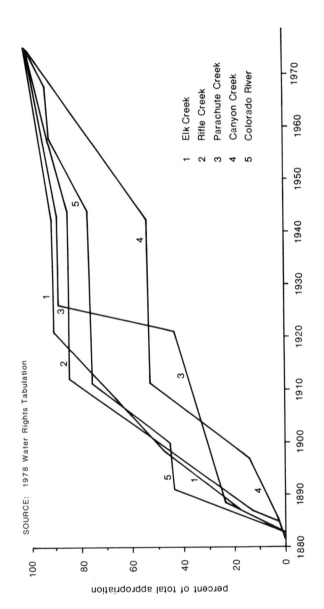

Fig. 9. Timing of absolute flow rights appropriations (1880-1970)

product of an historical surplus of water in the Upper Colorado River Basin. The surplus is also said to explain the paucity of wells and even of groundwater information in the study area.[10] Although the "surplus" argument may hold for the Colorado River itself, it will be shown that many tributary streams annually run dry in mid-summer, reducing hay yields in smaller drainages and in stream basins with inadequate storage of spring runoff. Two major reservoirs in the study area--Rifle Gap and Vega--were constructed in 1964 and 1968 as components of Bureau of Reclamation projects. A third smaller Federal project is planned for the West Divide area, and a major Colorado River reservoir is planned at Una. These proposed projects have encountered strong environmental and fiscal opposition. Energy firms and agricultural water districts are investigating a wide number of tributary storage projects, but few of these have gone beyond preliminary engineering studies.

The total magnitude of water rights in a basin depends in part upon physiographic factors influencing basin yield, and in part on the locational determinants of water demand. To compare the intensity of water rights appropriation in different basins, it is necessary to adjust for differences in basin area. Consequently, each decree total was divided by stream drainage area to establish a set of four "area-adjusted" decree types (AFLOW/Mi^2, CFLOW/Mi^2, ASTOR/Mi^2, and CSTOR/Mi^2, respectively).

The Colorado River represents a special case in the analysis. It is distinguished from its tributaries not only by its much larger drainage area, but also by the effects of storage reservoirs, transbasin diversions, and hydrologic phenomena occuring upstream from the study area. From the standpoint of water demand, the Colorado River establishes a major transportation corridor linking the most important economic centers in the region, again distinguishing it from the tributary regions.

The important point to be made here is that the Colorado River main stem and its many small tributaries have up until recently functioned as two nearly separate systems of surface water development. The main stem carries large volumes at relatively low elevations and often requires pumping for use, whereas the flashy tributaries support gravity flow diversion for limited volumes and

---

10. U.S. Department of the Interior, Geological Survey, *Summary Appraisals of the Nation's Groundwater Resouces - Upper Colorado Region*, Professional Paper, no. 813-C (Washington, D.C., 1974); idem, "Availability and Chemical Quality of Ground Water in the Crystal River and Cattle Creek Drainage Basins near Glenwood Springs, West-Central Colorado," Water Resources Investigation 76-70, open-file report (Lakewood, CO, 1976); and one of the few investigations in Garfield County--Wright Water Engineers, "Ground Water Resources in Garfield County, CO" (Glenwood Springs, CO, 1977). Interview with O. James Taylor, U.S.G.S., Denver, August 1981.

shorter times during the year. The water "surplus" of regional water policy-makers refers almost exclusively to the Colorado River main stem. In contrast, many tributary water users have faced scarcity and cutbacks on an annual basis, with appropriation in most slackening by the 1920's.

Until recently, these systems could operate independently without much consequence. Increasing demands placed on the main stem, however, could lead to administration of tributary use to meet senior main stem demands. Similarly, attempts to change a point of diversion from a tributary to the main stem will be met with increased opposition from main stem appropriators and transbasin diverters.[11] Consequently, when discussing the overall patterns of water rights appropriation, the Colorado River is included. When the discussion turns to local physiographic and economic conditions affecting appropriation, the Colorado River is considered an outlier and omitted from the analysis.

The relationships between storage and flow rights and between absolute and conditional rights will be considered before embarking on a description of individual decree types. Are these different types of water rights independent of each other? Table 2 would suggest not. Higher rates of direct flow appropriation, for example, should lead in time to increased storage development, which in turn should support still higher rates of direct flow diversion. The correlation coefficient of .7087 between absolute storage rights (ASTOR) and direct flow rights (AFLOW) strongly supports this expectation. Secondly, conditional flow rights should be concentrated in areas of emerging water demand and on larger streams whose supplies have not been fully developed. The latter factor would contribute to the positive correlation (.3967) between absolute and conditional flow rights (AFLOW and CFLOW) observed in table 2. Conditional flow rights do not appear to be geographically bound, however, to the location of existing reservoirs, whether because of full reservoir subscription, institutional constraints on marketing of reservoir supplies, or the cost of transporting water from existing reservoirs to the proposed place of use.[12] These interrelationships among types of water rights must be borne in mind in the interpretation of water rights patterns.

---

11. See Division 5 water court case no. W2127 in which a proposed transfer from Canyon Creek to the main stem was opposed by Denver.

12. Conditional storage rights are heavily concentrated in areas of proposed energy development, but the skewness and extreme kurtosis of their statistical distribution did not permit their inclusion in correlation models such as those given in table 2.

TABLE 2

CORRELATION COEFFICIENTS FOR DECREE TYPES (n = 48)

Variable	AFLOW	CFLOW
CFLOW	.3967 †	--
ASTOR	.7087 †	.0216

Variable	AFLOW/Mi^2	CFLOW/Mi^2
AFLOW/Mi^2	--	--
CFLOW/Mi^2	.3165 †	--
ASTOR/Mi^2	.1851	.0213

† α = .01; d.f. = n-2; two-tailed t test

How do local conditions affect the magnitude and intensity of water rights appropriation? To answer this question, the number and size of different decree types were compiled for all streams in the study area, including the Colorado River (Appendix, table 22). Aggregate figures were compared among the four administrative water districts. The Colorado River was then omitted so as to look more closely at tributary water rights patterns.

The analysis focused on the correlation between each of the four water rights types and selected physical, economic, and institutional variables. Listed below are the selected variables together with their expected relationships with water rights decrees.

1. *Physical and Hydrologic Variables*
    a) *Basin Area, AREA* (measured in square miles from U.S.G.S. 1:50,000 and 1:250,000 topographic sheets). Basin area was hypothesized to have a positive correlation with all decree types (before adjustment for area), and no association with area-adjusted decree variables.[13]

---

13. The distribution of area-adjusted and other ratio variables was observed to be approximately log-normal, and for this reason the natural log of these variables (e.g. AFLOW/Mi^2) was used in correlation models.

b) *Average Annual Precipitation, PRECIP* (measured in inches using the isohyetal method on Iorns' precipitation map, updated to reflect a longer period of record).[14] Precipitation was expected to be positively associated with both area-adjusted and unadjusted decree variables.

c) *Potential Basin Yield, YIELD* (calculated as AREA X PRECIP X 53.55; log transformation). The yield variable was hypothesized to have a strongly positive association with unadjusted decrees and a weakly positive association with area-adjusted decrees (due to the PRECIP variable).

d) *Stream Length, LENGTH* (measured in miles on U.S.G.S. 1:50,000 topographic sheets). Positive association with unadjusted decrees was expected due to more frequent opportunities for diversion on longer streams. No association was expected with area-adjusted decree variables.

e) *Channel Slope, SLOPE* (measured as the ratio of rise in elevation from the stream confluence to its head waters over stream length; log transformation). Negative associations were expected between SLOPE and unadjusted decrees since steeper slopes generate more rapid runoff. No association was expected between slope and area-adjusted decree variables.

2. *Location Variables*

a) *Distance, DISTANCE* (measured in miles from the stream confluence to the nearest municipality on a major road or rail spur). In keeping with the classical distance decay model, a negative association would be expected with all variables. Mitigating factors include, however, the importance of access to grazing areas and physiographic constraints on irrigation agriculture in different basins.

b) *Population, POPULATION* (from 1980 Census of Population for each of the municipalities mentioned in DISTANCE above). The fact that early settlements were often situated in proximity to the best water supplies was expected to result in a positive association between POPULATION and each decree variable.

---

14. W.V. Iorns, et al., *Water Resources of the Upper Colorado River Basin*, U.S. Geological Survey, Professional Paper, no. 441, (Washington, D.C., 1965). Records from 1921 to 1950 were updated to 1979, and earlier records were included for stations as available.

c) *Demand*, DEMAND (a composite measure of relative location obtined by dividing POPULATION by DISTANCE; log transformation). If the hypotheses regarding POPULATION and DISTANCE are correct, a positive association would be expected between DEMAND and decree variables.
3. *Basin Rank Variables*[15]
   a) *Appropriation Date*, YEAR (date of appropriation for the first decree in each basin; rank ordered to reflect the ordinal scale of the seniority system). Areas of earlier appropriation were hypothesized to be associated with more intensive levels of appropriation, reflecting the importance of reliable water supplies for early settlement.
   b) *Period of Continuous Appropriation*, PERIOD (beginning from the first date of appropriation and continuing until a lapse of ten years between decrees; measured in years). The period of sustained water appropriation was expected to be positively correlated with the decree variables, again reflecting the association between the timing of development and the overall intensity of development.[16]

The full set of correlation coefficients between decree variables (AFLOW, CFLOW, and so on) and the ten descriptive variables listed above is presented in the Appendix. Means for all variables are given in table 23. The most important findings are presented below, together with summary statistics and maps of water rights patterns for each of the four decree types.

Absolute Flow Rights

Comparison of absolute flow rights filings by basin revealed first, the importance of massive Grand Valley diversions from the Colorado River in district 72, and second, the much greater total drainage area of that district.

Compilation of absolute flow rights for each of the four water districts (table 3) showed that district 72 had not only the greatest number and magnitude of total appropriation but also the

---

15. An important question lies in whether any interaction in fact occurs among tributary basins, regardless of priority. Water commissioners suggested that little direct or indirect interaction has historically occurred in water rights administration, aside from cases of canal extension across basin boundaries and instances of direct upstream-downstream interaction. Further appropriation of the Colorado River main stem waters could stimulate more complex regulation of decrees among tributary basins.

16. It should be noted that whereas measures of water rights priority introduced in the following chapter are hypothesized to explain differences in water rights practice, here they serve a descriptive purpose in illustrating the correlation between the timing of appropriation and the eventual intensity of appropriation.

largest appropriations, as revealed in a mean decree size of 8.48 cfs, roughly three times that of other districts. District 70 had the smallest cumulative appropriation and the smallest number of AFLOW decrees, by contrast, but displayed an average decree size that was comparable with other districts.

TABLE 3

ABSOLUTE FLOW RIGHTS BY DISTRICT (AFLOW)

	Total No. of Streams n = 52	District 39 n = 8	District 45 n = 13	District 70 n = 8	District 72 n = 23
Sum of Decrees (c.f.s.)	14,531	2,102	1,557	742	10,130
Number of Decrees	2,960	708	764	294	1,194
Mean Size (c.f.s.)	4.91	2.97	2.04	2.52	8.48

NOTE: For location of the four water districts (39, 45, 70, and 72) and the 52 stream drainages, see figure 8.

The intensity of appropriation is defined here as the total magnitude of decree rights, adjusted for differences in drainage area (AFLOW/Mi^2). Appropriation of the Colorado River in district 72 clearly exceeds in intensity that of all other drainages (figure 10). When the Colorado River is excluded, however, it is district 45 which exhibits the most intensive development of absolute flow rights and district 70 the least intensive (table 4 and figure 10).

TABLE 4

INTENSITY OF ABSOLUTE FLOW RIGHTS (AFLOW/Mi^2)

	Total No. of Streams n = 48	District 39 n = 7	District 45 n = 12	District 70 n = 7	District 72 n = 22
Intensity (c.f.s./mi.^{2})	2.8	2.4	3.7	1.4	2.8

NOTE: Colorado River main stem excluded.

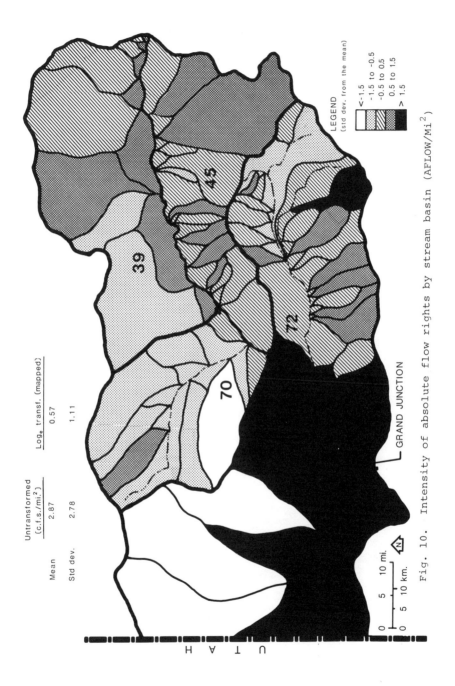

Fig. 10. Intensity of absolute flow rights by stream basin (AFLOW/Mi^2)

Why should this be so? First, local water managers state that district 45 basins have lower annual water yields than those in district 39 or 72. Second, drainages in district 45 are relatively small but have broad arable mesas and are well situated for irrigated pasture and forage crop production. Drainages in district 70, on the other hand, are large but relatively dry. Arable land is limited to constricted valleys, and streams are flashier than in other districts.

These explanations for differences in absolute flow rights may be examined further through measures of association with the physical, locational, and basin rank variables listed above. In district 45 correlation between physiographic variables (particularly stream length, drainage area, and basin yield) and absolute flow rights was highly significant, illustrating that in areas of relatively greater water demand, physiography plays an increasingly important role (Appendix, table 24). After adjustment for differences in drainage area, only precipitation is statistically significant (Appendix, table 25). Up to this point the hypothesized relationships have been fulfilled.

In district 70, however no physiographic variables obtained a significant level of association with absolute flow rights. Indeed the distance variable alone is significant, but in the opposite direction from that expected, indicating perhaps that proximity to the Colorado River corridor was less important than other location factors for forage crop production in this relatively remote district. As for the remaining two districts--district 39 displayed few significant correlation coefficients, whereas district 72 coefficients support the expected results for almost all variables.[17] Rather than pursue differences among districts in greater detail, it will be worthwhile to proceed to the remaining types of water rights, for the decree patterns affect each other in important ways.

Conditional Flow Rights

Compilation of conditional flow decrees revealed the importance of both future energy water demands (particularly in

---

17. Correlation between the physiographic variables and area-adjusted absolute flow rights went against the expectation of no association. When the area adjustment was made, the signs of correlation coefficients for physiographic variables were reversed, with the exception of precipitation. This may be explained by the fact that in district 72 the largest drainages also have the gentlest slopes, the longest stream lengths, and the lowest elevations. This last factor tends to result in lower average annual precipitation. Because the precipitation gradient is so strong in district 72, the signs of other physiographic variables could have reversed when the area adjustment was made. The precipitation variable would naturally remain unaffected. In short, the smallest basins are the most intensely appropriated because precipitation rates in them tend to be much higher.

district 39 which includes Parachute Creek) and of historic agricultural water demand in district 72 (table 5). Curiously, district 70, which is also slated for synfuels development, had the smallest number and total amount of conditional flow appropriations. Not surprising at all, however, was the relatively small mean size of conditional decrees in district 45 where streams are smaller and absolute flow rights more intensively developed than in other districts.

TABLE 5

CONDITIONAL FLOW RIGHTS BY DISTRICT (CFLOW)

	Total No. of Streams n = 52	District 39 n = 8	District 45 n = 13	District 70 n = 8	District 72 n = 23
Sum of Decrees (c.f.s.)	11,089	5,306	1,388	853	3,452
Number of Decrees	526	113	173	34	206
Mean Size (c.f.s.)	21.08	46.96	8.02	25.09	17.19

After adjusting for differences in drainage area, the pattern of appropriation reveals a clear concentration along the Colorado in districts 72, 70, and particularly 39 (figure 11). Low yielding tributaries in district 70 have the least intensive conditional appropriation, in spite of their proximity to energy development sites (table 6). Tributaries with reservoir development potential in district 39 such as Elk Creek also have above average rates of conditional direct flow appropriation. In district 45, new town and exurban development, in conjunction with proposed canal extensions, account for areas of more intensive conditional appropriation.

TABLE 6

INTENSITY OF CONDITIONAL FLOW RIGHTS (CFLOW/Mi^2)

	Total No. of Streams n = 48	District 39 n = 7	District 45 n = 12	District 70 n = 7	District 72 n = 22
Intensity (c.f.s./mi.^{2})	1.2	2.5	2.3	0.4	0.6

NOTE: Colorado River main stem excluded.

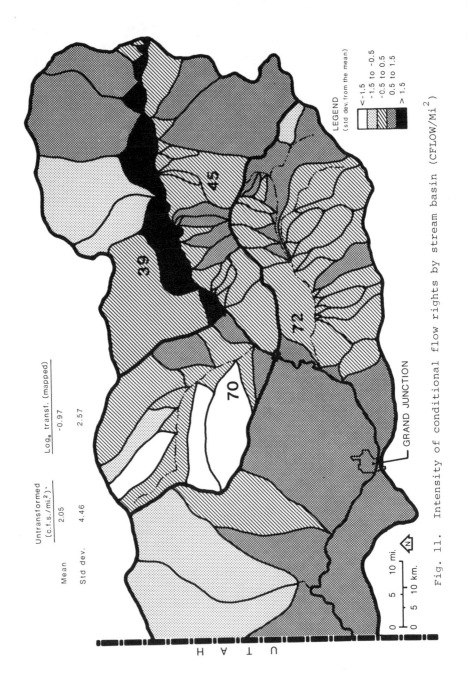

Fig. 11. Intensity of conditional flow rights by stream basin (CFLOW/Mi²)

Conditional flow decrees displayed similar though generally weaker associations with physiographic and economic variables than was the case with absolute flow rights (Appendix, table 26). The most important difference occurred for the limited tributary supplies of district 70 where physiographic variables took on greater importance than was apparent for absolute flow rights. Physiographic variables were again strongly significant for districts 45 and 72. After adjusting for differences in drainage area, precipitation emerged as the most consistently important factor associated with conditional decrees (Appendix, table 27).

Absolute Storage Rights

The volume of absolute reservoir storage rights varies by more than two orders of magnitude among the four water districts (table 7 and figure 12). Storage for the Collbran project and Grand Valley irrigation area in district 72 account for the largest decrees, followed by the Silt Project in district 39. The absence of major Federal storage projects in districts 45 and 70 accounts for their proportionately smaller storage decrees. Even after adjustment for differences in drainage area, approximately the same relationships are found among tributary storage volumes (table 8). Correlation of storage rights volumes with physiographic and location variables was strong in district 72, and to a lesser extent, in district 45 (Appendix, tables 28 and 29). These results stem from the fact that storage rights are less developed in areas of western Colorado not benefitted by the Federal reclamation program.

TABLE 7

ABSOLUTE STORAGE RIGHTS BY DISTRICT (ASTOR)

	Total No. of Streams n = 52	District 39 n = 8	District 45 n = 13	District 70 n = 8	District 72 n = 23
Volume of Decrees (acre-feet)	86,875	22,816	2,096	524	61,439

Conditional Storage Rights

Conditional reservoir storage decrees are heavily concentrated in several basins and completely absent from many others (figure 13). Whereas absolute storage decrees are concentrated on the larger tributaries, conditional storage decrees are greatest on the

TABLE 8

INTENSITY OF ABSOLUTE STORAGE RIGHTS (ASTOR/Mi^2)

	Total No. of Streams n = 48	District 39 n = 7	District 45 n = 12	District 70 n = 7	District 72 n = 22
Intensity (a.f./mi.^{2})	24.3	23.0	5.8	1.6	42.1

NOTE: Colorado River main stem excluded.

Colorado River main stem, particularly in districts 39 and 70 (tables 9 and 10). Storage decrees for the Colorado River account for roughly sixty percent of total conditional storage volumes. Larger tributaries such as Canyon, Elk, Parachute, Divide, and Roan Creeks account for the greater part of the remainder. More widespread conditional storage rights in the upper Plateau Creek drainage reflect the importance of higher precipitation rates as well as downstream demand in the Grand Valley area. Conditional storage rights, of all the decree types discussed, most clearly reflect the emerging pattern of water demand--first, for energy development, and second, for associated municipal development. The water rights strategies of these most recent uses are necessarily complex, however, and deserve explicit treatment.

TABLE 9

CONDITIONAL STORAGE RIGHTS BY DISTRICT

	Total No. of Streams n = 52	District 39 n = 8	District 45 n = 13	District 70 n = 8	District 72 n = 23
Volume of Decrees (CFLOW)	969,230*	412,786	24,262	397,644	134,538

*Includes double counting of adjudications for Una Reservoir in districts 39 and 70 (decree volume = 195,983 acre-feet).

Water Rights for Oil Shale Development

Much of the concern expressed over water supplies is directly attributable to the emergence of a synfuels industry in the region. Although the tide of opinion as to how these new demands will be met has recently shifted from concern to complacency, it is still not

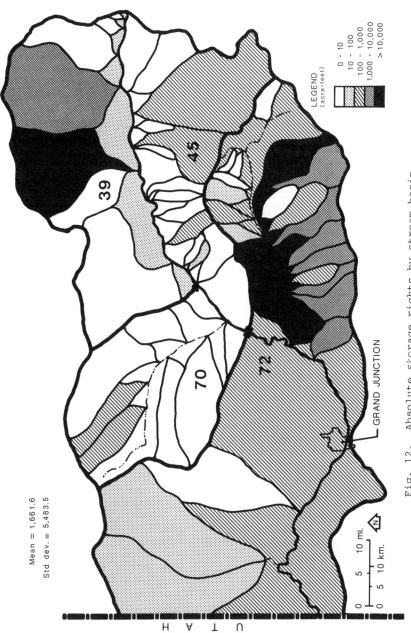

Fig. 12. Absolute storage rights by stream basin

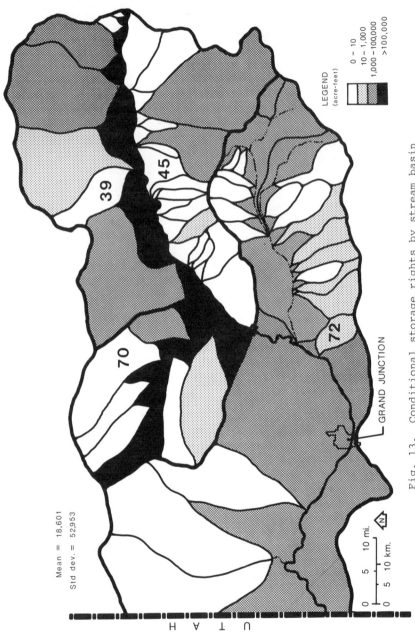

Fig. 13. Conditional storage rights by stream basin

TABLE 10

INTENSITY OF CONDITIONAL STORAGE RIGHTS (CSTOR/Mi^2)

	Total No. of Streams n = 48	District 39 n = 7	District 45 n = 12	District 70 n = 7	District 72 n = 22
Intensity (a.f./mi.^{2})	93.0	228.5	18.4	40.6	107.1

NOTE: Colorado River main stem excluded.

clear how the numerous energy-related water rights filings and changes will affect the water resource system. Part of the problem lies in the absence of water rights ownership information in Colorado. Even if such information were available, however, the purported water rights options, leases, and shareholdings of energy firms and residential developments would remain unknown. Nonetheless, several types of information provided a glimpse into the water rights holdings and acquisition strategies of energy firms: project representatives were interviewed, land holdings reviewed, conditional decrees examined, and diversion structure ownership records investigated. Taken together, this information gives a balanced view of the fabric and patterning, if not the exact proportions of water rights appropriation for industrial and municipal growth.

Shale oil project representatives willingly discussed their overall water acquisition strategies, but only in several cases supported these discussions with detailed references to current water rights holdings.[18] In each case a previous summary of the firm's water rights holdings, compiled by the University of Wisconsin Water Resources Institute, was presented to the interviewee. Confirmation or correction was requested. The principal goal expressed by most energy representatives was to obtain secure title to a quantity of water that would meet initial project needs in all years. Secondary supplies were also sought for future projects and production scale-ups.

Given the uncertainty of water rights yields and project requirements, most firms have assembled a portfolio of rights that includes senior irrigation rights, conditional direct flow decrees

---

[18]. The author's fellowship with the Tosco Foundation quite possibly opened channels within the oil shale firm of that name while restricting communications with some competing firms.

from the Colorado River, conditional storage rights on tributaries, well permits, and options for reservoir storage water at existing or proposed Federal reservoirs. Acquisition of senior irrigation rights was downplayed as politically unpopular and as inefficient (because of high transaction costs and reduction of the transferable portion of the acquired right), though twenty years ago a different view held sway. Irrigation rights that have been acquired are usually leased back to the farmer for taxes, and presently no firms claim to be actively seeking these rights. To summarize, it appears that the initial industrial water rights demand surge has been more than met and that firms are looking to choose among water rights options they now hold.

The location of energy water rights filings closely follows project locations, conveyance requirements, and physiographic opportunities for reservoir siting (figure 14). Although oil shale projects and water rights are concentrated in three tributary basins, water rights holdings extend throughout water divisions 5 and 6. Of all the lands held in shale leases, only one project is actively proceeding at present--a situation characteristic of the fitful synfuels industry.[19] Earlier estimates in the 1970's of a 2 million bbl/day industry by the turn of the century justifiably stirred up concern over water supplies, as did talk of water importation from Oahe Reservoir on the Missouri River. The pace of industrial development has not met these expectations, however, so that more careful consideration of water development opportunities should be possible. While few projects have started, many rights have been acquired not only by energy firms but also by land developers, who by county regulation must give evidence of a firm water supply as a condition for subdivision and zoning approval.[20] Subdivision plans are myriad, but mortgage financing rates and lending rates have forestalled a boom in housing construction and postponed the widespread water rights transfers characteristic of the Front Range metropolitan corridor from Fort Collins to Pueblo.[21] Condemnation of agricultural water rights is possible but has not yet occurred in municipal areas. Municipal purchases of water rights on annexed or upstream lands have thus far been adequate. More detailed analysis of land use changes lies outside the scope of this study and in the end would not provide complete information on

---

19. Colony, one of the most advanced of these projects, was recently postponed. *Wall Street Journal,* May 4, 1982.

20. Interview with Ray Baldwin, Garfield County Energy Impact Coordinator, Glenwood Springs, July 1981.

21. Raymond L. Anderson and Norman I. Wengert, "Developing Competition for Water in the Urbanizing Areas of Colorado," *Water Resources Bulletin* 13 (1977): 769-773.

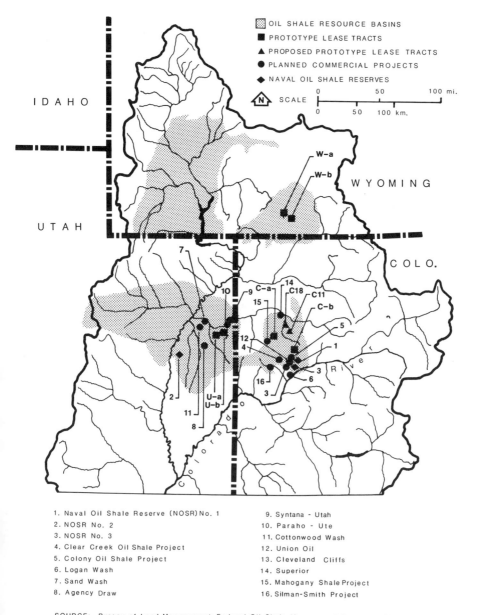

Fig. 14. Map of oil shale project lands in the study area

water rights holdings because the two sets of property rights are separable on all but Federal reclamation project lands.

The water rights tabulation includes conditional decrees which provide some insight into water demand for industrial use. A 1975 study by the University of Wiscosin compiled these for each energy firm and mapped their distribution. The Colorado Water Conservation Board then revised these findings using the water rights tabulation and a survey of energy firms. Further revisions were made for this study using the water rights tabulation, structure ownership information, and interview data. The structure ownership survey revealed that the acquisition of senior irrigation rights has with few exceptions been highly localized near project areas. Exceptions include water rights acquired in district 45 for Battlement Mesa, Inc. new town and Union Oil's experiment in water transfers from the Roaring Fork drainage. It should be noted that structure ownership files indicate only a single owner, whereas several owners or shareholders may exist.

Conditional decrees compiled from the water rights tabulation reveal massive appropriations that in many cases will never be developed. This may raise again the issue of what constitutes speculative appropriation and reasonable diligence. For Colony, Chevron, Union, and Mobil the intent to appropriate was demonstrated by the start of project construction--dormant projects are another matter. Should the holders of these rights be able to sell them for use at another site without loss of the original priority date? How often should surveys, legal studies, and other non-construction activities suffice in findings of due diligence? Some analysts have suggested that the late date of appropriations (1950's and after) make these issues moot, but the combined diversion and storage rights are by no means without value in most years and are in all years superior to subsequent decrees.

Although several firms have had options for Federal reservoir water, actual contracts for water are very recent in date and limited to Reudi Reservoir of the Frying Pan-Arkansas Project. The first three contracts for Reudi water were concluded in Fall 1981 after more than a year of serious negotiation. Contracts were let to the Colony oil shale project including the Battlement Mesa new town, the Basalt Water Conservancy District, and the West Divide Water Conservancy District. Historically, the Bureau of Reclamation (USBR) water has been sold at subsidized prices. Initially USBR wanted to price Reudi water based not on a repayment schedule for the dam, but on the marginal costs of water for industrial use, a sum they initially calculated at approximately $155 per acre-foot.

The Federal agency also wanted to employ its risk-sharing policy in the event of a drought, whereby all contracts suffer proportionately. Colony sought to obtain historical prices for water, roughly $15 per acre-foot, and a priority system of curtailment from which they would obviously benefit in the event of drought. The compromise eventually reached stipulated graduated rates for delivered water and standby water, and a modified priority system for allocation under conditions of scarcity (table 11). The negotiated prices are less than most other water acquisition alternatives, such as acquisition of senior irrigation rights and construction of tributary storage reservoirs. In most cases diversion and reservoir facilities will still have to be constructed to meet project requirements, but not at the scale or capacity originally foreseen.

TABLE 11

PRICES FOR REUDI RESERVOIR WATER

Type of Use	Quantity (a.f./yr.)	Standby* Charge (per a.f.)	Delivery Schedule	
			Quantity (a.f.)	Price (per a.f.)
Industrial				
Colony Project	6000	$15	0-2000	$40**
			2000-4000	$60
			4000-6000	$80
Municipal				
Battlement Mesa West Divide WCD Basalt WCD	1250	$ 6	0-1250	$ 9***

SOURCE: Draft Contracts from U.S. Bureau of Reclamation, Fryingpan - Arkansas Project, Colorado (9/28/81).

*In the event of shortage, all contracts are reduced proportionately up to a maximum of 30%, afterwhich they are fulfilled in order of priority by contract date.

**In addition to annual standby charge. Minimum delivered quantities are stipulated throughout the contract life (until 2019).

***Adjustable upwards, if necessary, on a schedule of maximum total charges throughout the contract life (until 2019), but never to exceed $55/a.f. including standby charges.

Water rights acquisition for energy development, with the exception of Federal project water, seems to have become a subject for retrospective evaluation rather than forecasting or prediction. Forecasting efforts confirm the adequacy of existing holdings for the initial wave or waves of shale oil development, a fact supported by the slowed pace of water rights trading and development by firms. This is not to say that water rights patterns have stabilized, rather they are now tied to the land market and the secondary effects of energy development on population, housing, and recreational development. Those changes in water demand are more numerous and even more difficult to survey than energy water demands.

Summary

Water rights appropriation occurred most intensely during the period from 1880 through 1920. Subsequent actions have included water rights changes, large-scale conditional appropriation, and, occasionally, storage rights filings. The Colorado River main stem has had continued appropriation and represents an administratively, if not hydrologically, separate system of water rights. It is clear from water court cases that conflicts between the main stem and tributaries will increase as conditional decrees are perfected and absolute rights are changed for energy development. Increased interaction among tributary basins may also occur as canal consolidation, water rights conflicts, and administrative regulation become more frequent.

In the description of tributary water rights patterns, it was observed that physiographic variables showed more widespread and stronger measures of association with decree variables than either location or basin rank variables. Precipitation often remained significant and stable after adjustment for basin area, particularly where a steep precipitation gradient exists, as in district 72. Basin rank variables paralleled absolute decree volumes in areas such as district 45, giving support to the idea that the best water supplies were appropriated first and allowed the longest duration of water rights appropriation. At this local scale only conditional water rights seem closely tied to specific locational conditions, that is, to oil shale development in districts 39 and 70. Both absolute and conditional storage decrees, on the other hand, clearly reflected physiographic requirements.

Water rights patterns provide only a partial foundation for interpreting water use and management. They become of crucial importance, however, in understanding the distribution and strength of participants in the water development process. In order to fully grasp the prospects for integrated water development and the real, as opposed to nominal, strength of vested water rights, water rights patterns must be compared with actual patterns of water use.

CHAPTER V

PATTERNS OF WATER DIVERSION

In this chapter the areal distribution of regional water supplies is presented in greater detail through an evaluation of district water commissioners' annual diversion records. Water diversion patterns frequently vary from water rights patterns. Water rights patterns, as already noted, represent a cumulative legal history of water development and must not be confused with what actually happens "in the field." Maass and Anderson caution against making inferences about water use that are derived from water rights analysis:

> One gains a highly imperfect sense of how water is distributed in irrigated areas by reading the vast and prolix literature on the legal nature of water rights. There is a difference between legal concepts of water rights and water practice, and many students of irrigation have overstressed the importance of rights about which they can write at length without leaving their desks.1

Their critique stops somewhat short, however, by passing over the causes and impacts of discrepancies between legal and actual patterns of water management. As Elwood Mead observed, discrepancies arose as a result of exaggerated filings and inadequate field checking in the adjudication process. Resulting problems have occured in field administration of adjudicated decrees, in recording the use of water rights, and in establishing beneficial use and injury in court proceedings. To be sure, precise legal description of actual water use patterns would require platoons of recorders and vast data handling systems.

Although the critique of water rights investigations is well aimed, attention to these problems and to the consequent evolution in water practice and water rights systems is necessary. The following discussion begins with a survey of water use records and

---

1. Arthur Maass and Raymond L. Anderson, *. . . and the Desert Shall Rejoice* (Cambridge, MA: MIT Press, 1978), p. 424; see also Michael V. McIntire, "The Disparity Between State Water Rights Records and Actual Water Use Patterns," *Land and Water Law Review* 5 (1970): 23-48.

record-keeping practices. Water diversion patterns are then contrasted with water rights patterns to assess the magnitude of the "paper rights" problem. The remainder of the chapter provides a detailed description of water diversion patterns in the study area.

Public Use Water Records

The office of water commissioner was created in the Irrigation Act of 1879. Its present duties have evolved to consist of recording diversions; supervising water exchanges; checking and setting headgates; cutting off junior appropriators and ungauged structures in times of shortage; and making field determinations of "futile calls," waste, and water theft. Although the 1969 Water Rights Determination and Administration Act extended administrative duties to include more issues of public interest, it must be emphasized that water administration serves principally to protect property rights to water in accordance with the prior appropriation doctrine. Protection of property rights comes in two forms--first in actual field regulation of diversions during water shortages. It is a common misconception that only senior appropriators benefit from field administration of water diversions. On the contrary, even the most junior appropriator suffering a shortage may request a call on the stream and may benefit by senior appropriators being cut back to their decreed flow rights.[2] The second form of property rights protection lies in the legal record of water withdrawals. This record can have great importance for determination of historic beneficial use which is, in essence, the appropriator's transferable property right.

By law, any appropriator is entitled to timely measurement of decrees affecting his property right, and failure of public officials to do so constitutes a misdemeanor. (C.R.S. 1973, 37-84-122). Prior to 1950 records included detailed information on crops grown, acreage irrigated, and field notes by commissioners and deputy commissioners. Today annual measurements generally include the following information:

1. Number of days water is diverted
2. Average rate of flow taken through the structure (c.f.s.)
3. Volume taken (acre-feet)
4. Acres irrigated (if applicable)
5. Irrigation application rate (acre-feet per acre)

---

[2]. Personal communication with Lee Enewold, Division 5 Engineer, and water commissioners.

6. Exchanges, loans or transfers, if any
7. Reasons for omissions in the record, for example, no measuring device, no water taken, or no information available.

The last reason for records not being taken, "no information available," deserves emphasis from the outset. Very few diversions in Division 5 are measured continuously (e.g. Grand Valley and Shoshone Dam), and some are never recorded. Many records arise only by request or in the resolution of specific conflicts rather than in the systematic survey of water use.

Limitations of Diversion Records

Diversion records in Division 5 are sketchy and vary in accuracy, often being based on a single observation which is then extrapolated to obtain annual totals. The recorded information is often supplemented by informal input from the property right owner himself regarding the number of days water was taken or the number of acres irrigated. Prior to 1970 many structures had neither headgates nor measuring devices, as required for proper administration (C.R.S. 1973, 37-84-112).[3] Even following the installation of these structures, 1980 diversion records show a majority of decreed diversion structures as still having "no record."

Excluding sedimentation and washouts of diversion structures in localized areas, the main reason for poor diversion records stems from inadequate staffing to meet both the needs of the area and the legislated responsibilities assigned to the Division of Water Resources. District water commissioners may be responsible for administering literally thousands of rights in several counties. Neither wells nor well logs are commonly checked. Reservoir records come from reservoir operators and not water commissioners.

The historical "surplus" and fiscal constraints again receive the blame for loose administration in Water Division 5. Budget requests for 1982 included provisions for more water commissioners and deputies in recognition of increased competition and conflict over water in the region.[4] These problems extend far beyond the initial diversion of water from a public stream. Most local water organizations have also encountered increased conflict in water

---

3. The Division 5 Engineer's Annual Report for 1971 states that headgate notices were issued to all structure owners who had not yet installed such devices. The 1974 and 1975 reports contain a resolution on the need for installation of measurement devices to properly administer water diversions in the region.

4. Interview with D. Monte Pascoe, Director, Colorado Department of Natural Resources, 1981.

distribution among their users, particularly in areas experiencing rapid suburbanization or subdivision.[5]

Patterns of Record-Keeping

The problem of missing records, in spite of posing important qualifications on the findings made, offers an opportunity to explore the pattern of water administration and the distribution of conflict in each district. Patterns of record-keeping have both methodological and substantive importance. Water diversion records are in effect a sample survey of water rights utilization. If this sample is systematically biased, then the resulting interpretation of diversion data should attempt to identify and evaluate the effect of those distortions on the findings obtained.

A major theme in this chapter has been concerned with what field administration records suggest about actual water practice. From that standpoint, the patterns and biases of administrative activity should yield more than simply methodological tidiness when subjected to investigation. What factors explain the thoroughness of administrative record-keeping? In water Division 5, officials like to say that "the squeaky wheel gets the grease" meaning that areas of conflict receive the closest administrative attention. Second, they cite the reluctance of some water users to install and maintain the water control structures necessary for accurate record-keeping, particularly for less valuable water rights. Inadequate staffing is also said to preclude thorough record-keeping. General hypotheses thus include first, the expectation that the more valuable water rights, that is, the larger and more senior rights, receive the most administrative attention; second, that "hot spots" of chronic conflict and water shortage also receive more detailed administration; and third, that administrative resources and responsibilities are not distributed in the most efficient or effective manner.

The proportion of water diversions annually recorded in individual drainages ranges from 0 to 100 percent of all structures (figure 15). In the following watersheds no diversions were recorded in 1980:
1. District 45--Porcupine Creek and Spring Creek
2. District 72--Big Salt Creek, East Salt Creek, West Salt Creek, Owens Creek, Collier Creek, Cheney Creek, and Yule Creek.

---

5. This issue will be taken up in greater detail in chapter eight.

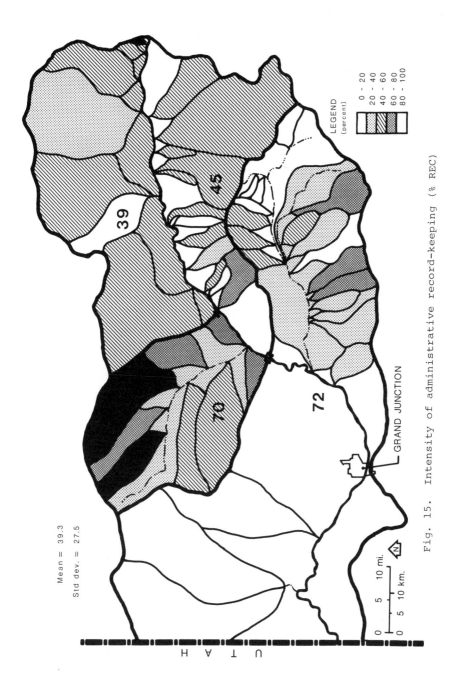

Fig. 15. Intensity of administrative record-keeping (% REC)

District 70 had the greatest percentage of structures recorded (67.7 percent), while district 72 had the lowest (41 percent) (Appendix, table 30). In most cases the unrecorded structures represent small volumes of appropriated water, averaging less than 5 cfs per structure. Annual summary records do not indicate the frequency of measurement at individual structures, however, and it could be that in district 72 fewer structures are administered more intensively.[6] Nevertheless, the "squeaky wheel" hypothesis is clearly not borne out in all cases. Porcupine Creek, a perennial hot spot in district 45, had no diversion records for 1980. Furthermore, the relatively complete record set in district 70 stands in contrast with the water commissioner's statement that little conflict exists anywhere in that district over the administration of water rights.[7]

From conversations with water officials, it is clear that diversion records are not a random sample of water rights or of water users. What these records do in fact represent may be probed further by using measures of association between the percentage of structures recorded and the following stream basin variables which were fully described in chapter four:[8]

1. *Physical and Hydrologic Variables*
    a) Stream Length, LENGTH
    b) Basin Area, AREA
    c) Average Annual Precipitation, PRECIP
    d) Potential Basin Yield, YIELD
2. *Location Variables*
    a) Distance, DISTANCE
    b) Demand, DEMAND
3. *Basin Rank Variables*
    a) Appropriation Date, YEAR
    b) Period of Continuous Appropriation, PERIOD.

Few of these variables appeared to have any important association with the percentage of structures recorded (Appendix, table 31). In some districts larger basins and longer streams had a smaller percentage of structures recorded. In others record-keeping appeared to be concentrated at greater distances from older, more settled areas, suggesting that conflict might be more prevalent in

---

6. This information could be obtained by going beyond the annual summaries employed here to the complete records of diversion for each structure.

7. Conversation with George Anderson, district 70 water commissioner, in DeBeque, Colorado, 1981.

8. For correlation models, a log transformation was used to normalize the sample distribution of the "percentage of structures recorded" (% REC) variable.

such areas. Otherwise, physiographic and location variables appear to have little association with administrative practice.

More interesting results emerged from an examination of relationships between the percentage of structures recorded (% REC) and other water diversion variables. The following water diversion variables were compiled from the 1980 water commissioners' summaries and the water rights tabulation:

1. *Total Water Diverted (CFS),* measured in cubic feet per second for each stream
2. *Area-adjusted Diversions (CFS/Mi.),* total water diverted divided by basin area (cubic feet per second per square mile; log transformation)
3. *Irrigation Application Rate (AF/AC),* total water diverted divided by basin area (acre-feet per acre; log transformation)
4. *Length of Water Diversion Season (DAYS),* number of days per year that water is diverted
5. *Total Acres Irrigated (ACRE),* listed by diversion structure and compiled for each stream basin
6. *Mean Rank of Recorded Structures (MRANK),* relative priority of structures recorded (individual ranks were averaged for each stream then rank ordered for correlation).

Positive correlation coefficients between the percentage of structures recorded (% REC) and both irrigation application rates (AF/AC) and seasonal length (DAYS) indicate that the longest and heaviest irrigators receive the most administrative attention. These results suggest that in many areas owners of more valuable rights are not simply complying with administrative requirements but rather are demanding protection of their property rights.

If record-keeping were simply a function of conflict on an annual basis, it would not be biased toward the more valuable rights. As the Division Engineer pointed out, conflict is in some cases more intense among junior appropriators operating at the margin of water availability. A legal record of water use provides administrative protection against future charges of abandonment or non-use. If sale of the water right is contemplated, users would particularly want to document the largest volume of historic beneficial use possible in order to maximize the transferable portion of their property right.

Comparison of Water Rights and Water Diversion Patterns

How closely do actual patterns of water diversion reflect the water rights patterns described in chapter four? The correlation between total water diverted in 1980 (CFS) and the sum of all absolute flow rights (AFLOW) was very high for all districts in the study area (table 12). After adjusting these diversion and water rights variables for differences in stream drainage area (CFS/Mi^2 and AFLOW/Mi^2, respectively), correlation results were again highly significant. There was a much lower, but still positive correlation coefficient between diversions (CFS) and absolute storage rights (ASTOR). All of this suggests that there is a correspondence between legal and actual patterns of water development.

When the total volumes of diversions recorded are compared with total water rights volumes, however, Mead's critique of water practice in Colorado reemerges (table 13). Even after correcting for unrecorded diversions, it becomes evident that in all districts actual diversions average only 29 to 52 percent of absolute flow rights. In chapter six it will be shown that 1980 diversion records are not uncharacteristic; the paper rights problem does indeed exist and acts to hinder rational water management in western Colorado.

Water Diversion Rates

Recognizing that water diversion records are somewhat biased toward more valuable water rights and that they represent only a partial sample of water use, they are still useful for describing broad patterns of water diversion. What geographical factors aid in understanding variability in water diversion pratices? Our approach to this question involves the comparison of district diversion variables with the same physical, locational, and basin-rank variables considered in the water rights analysis. These are listed below.

Water Diversion Variables
1. Total Water Diverted (CFS)
2. Area-Adjusted Diversions (CFS/Mi^2)
3. Irrigation Application Rate (AF/AC)
4. Length of Diversion Season (DAYS)
5. Total Acres Irrigated (ACRE)
6. Mean Rank of Recorded Structures (MRANK)

Stream Basin Variables
1. LENGTH
2. AREA
3. PRECIP

TABLE 12

CORRELATION BETWEEN WATER RIGHTS AND DIVERSION VARIABLES

Correlation Model	Total n = 39	39 n = 7	45 n = 10	70 n = 7	72 n = 15
TCFS by AFLOW	.8157**	.7676*	.9400**	.6907+	.8345**
CFS/Mi² by AFLOW/Mi²	.8049**	.8726*	.8192**	.7736*	.8311**
TCFS by ASTOR	.3736*	.5404	.3233	-.2562	.3428

+ $\alpha = .10$ * $\alpha = .05$ ** $\alpha = .01$; d.f. = n-2; two-tailed t test

Where: TCFS is the sum of annual recorded diversions
AFLOW is the sum of absolute flow rights
CFS/Mi² is area-adjusted diversion rate
AFLOW/Mi² is the area-adjusted sum of absolute flow rights
ASTOR is the sum of absolute storage rights

NOTE: Colorado River main stem excluded from each district.

TABLE 13

COMPARISON OF TOTAL DIVERSIONS AND TOTAL ABSOLUTE FLOW RIGHTS

Variable	Total n = 43	Total n = 39	39 n = 8	45 n = 11	70 n = 8	72 n = 16
Total Absolute Flow Rights (AFLOW)	14,233	5,772	2,024.8	1,518	741.6	9,950.4
Total Diversions (TCFS)	3,455.4	1,524.9	439.4	421.4	299.7	2,294.9
Total Diversions plus Volume of unrecorded Decrees* (TCFS + VNREC)	4,569.1	2,301.0	670.6	709.6	382.9	2,806.9
Ratio of Diversions to Decrees	.321	.399	.331	.467	.516	.282

*The volume of unrecorded decrees was estimated by summing the decreed volume for structures listed as having "no record" in the water commissioners' reports. Unlisted structures were not included, though for the most part they carry smaller flows; inclusion of the full decree amount for "no record" structures compensates for these omissions.

4. YIELD
5. DISTANCE
6. DEMAND
7. YEAR
8. PERIOD
9. POPULATION

As with the water rights records, it was hypothesized that larger water supplies (measured in terms of basin yield, average precipitation, and stream length) would be associated with larger diversions, larger irrigation application rates, longer irrigation seasons, and larger irrigated acreages. Larger diversions were expected to be closely associated with areas of larger and earlier settlement (i.e., positively with POPULATION, DEMAND and PERIOD, and negatively with DISTANCE and YEAR). Quite clearly, the diversion variables should also be correlated among themselves. An important question concerns the relationships between seniority and water diversion practice. As working hypotheses, one would expect that more senior irrigators make larger diversions (CFS, CFS/Mi^2), longer diversions (DAYS), and more wasteful applications (AF/AC).

Size of Diversions

Local differences in the total rate of water diversion (CFS) parallel differences in total water rights appropriation. District 72 displays a rate of diversion nearly an order of magnitude greater than district 70, which has the lowest total diversion rate (Appendix, table 30). Presumably, these results are at least in part a function of differences in drainage area. Highly significant correlation coefficients between diversion rates and physiographic variables confirm the importance of drainage area in three water districts and for the study area as a whole (Appendix, table 32). Only in the tributaries of district 72 does precipitation dominate area-based variables, probably as a result of two factors: low relative variability in tributary basin size and a full gradient of precipitation rates in the Plateau Creek drainage.

After correcting for differences in drainage area (CFS/Mi^2), it is the water-scarce district 45, and not district 72, which has the highest rate of diversion (Appendix, table 30). Water scarcity in district 45 is thus as much a function of greater water demand as it is a lack of adequate storage, as is commonly argued. The map of area-adjusted diversion rates (figure 16) illustrates the significance of the following factors on water diversion rates: reservoir storage in districts 39, 45, and 72; high precipitation

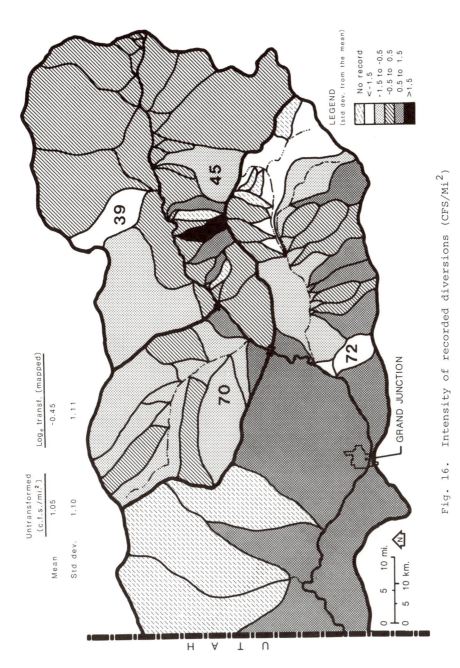

Fig. 16. Intensity of recorded diversions (CFS/Mi²)

rates in the Plateau Creek drainage of district 72; intensive water demand in the Grand Valley area and to a lesser extent in district 45. The entire Roan Creek drainage in district 70 and the Parachute Creek drainage in district 39 display the lowest rates of water diversion--a fact which illustrates low historical demand in areas slated for energy development.

As was the case with area-adjusted water rights, precipitation remained positively correlated with area-adjusted diversion rates, while other physiographic variables changed in sign (Appendix, table 33). An interesting divergence from water rights patterns is observed, however, in the strength of population and demand variables. Whereas water rights variables exhibited no significant relationships with these location variables, analysis of diversion rates shows that *less* intensive diversion occurs near the larger settlements. This finding illustrates first, that larger settlements are located in proximity to larger water supplies; second, that water use intensity is often greatest in basins with reliable yet relatively small supplies, such as in district 45; and third, that actual patterns of water use reflect the importance of relative location variables to a greater extent than do water rights patterns.

Scale of Irrigation

Irrigated acreage is concentrated in the Grand Valley area and in the Colorado River drainage generally (Appendix, table 30). Among the tributary basins, the mesas of district 45 are most extensively irrigated, followed by canyon bottoms of districts 39 and 70. A consistent pattern of positive association was observed between the irrigated acreage variable (ACRE) and physiographic variables such as stream length, basin area, and basin yield (Appendix, table 34). Whereas more intensive water diversion tends to occur at greater distances from population centers, irrigated acreage is greater in areas close to population centers. As would be expected, large scale irrigation is positively correlated with total diversion rates and water rights appropriation (CFS and AFLOW, respectively).

Length of Water Diversion Season

The argument that water availability varies widely among small watersheds is well supported by differences in the seasonal length of diversions (Appendix, table 30). In district 45 the seasonal length of water withdrawal averages only 76 days per year, supporting the claim of water scarcity there, whereas in every other district it averages over 100 days per year. When streams are considered individually, the shortest irrigation season runs only 39 days in Dry Hollow (district 45). The longest irrigation season runs 274 days on the Colorado River in district 72 to supply the Grand Valley area.

Association between the length of seasonal water diversion (DAYS) and stream basin variables diverges sharply from the results obtained with other diversion variables (Appendix, table 35). Physiographic variables, water rights variables, and water diversion rates have surprisingly weak relationships with the length of diversion. Instead, the rank variables (i.e., MRANK and YEAR) have greater importance, reinforcing the view that senior users have the most reliable and secure water supplies. Further support for this finding comes from the observation that longer diversions occur in relatively closer proximity to older settled areas.

Irrigation Application Rates

Irrigation efficiency is one of the most critical issues in this research. Excess water withdrawn from the stream can result in waterlogging, nutrient losses, saline return flows, and drainage problems; and it often precludes timely water use by junior appropriators. Irrigation efficiency will be discussed in greater detail in chapter seven, but diversion records introduce the subject by noting the rate of water diverted per acre of irrigated land. In some areas irrigation application rates do not even meet crop requirements, whereas in others they greatly exceed them.

In 1980 diversion rates for irrigation ranged from an average basin low of 2.91 acre-feet per acre in district 45 to over 13.0 acre-feet per acre in district 70 (Appendix, table 30). Overall district diversion rates for irrigation were less wide-ranging but followed a similar pattern. The map of irrigation application rates (figure 17) illustrates that the heaviest withdrawals occur in districts 39 and 70, that is, in those areas which often had the least intensive diversions (cf. figure 16 on CFS/Mi^2). District 45, which had the most intensive diversions and the shortest irrigation seasons, predictably has the lowest irrigation application rates. At the individual stream basin level mean irrigation rates vary from 0.90 acre-feet per acre in the Government Creek drainage (district 39) to 31.32 acre-feet per acre in Carr Creek (district 70). At the finest level of analysis--that of individual diversion records--one Carr Creek record totals an astounding 88.73 acre-feet per acre while a Beaver Creek (45) record

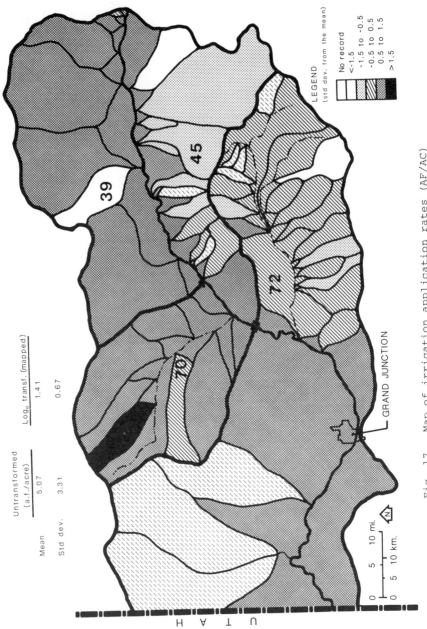

Fig. 17. Map of irrigation application rates (AF/AC)

indicates an irrigation application rate of only .08 acre-feet per acre.

How is it possible to have such variability and such apparent excess? Few of the descriptive stream basin variables offer much insight on variation in irrigation application rates (Appendix, table 36). Physiographic and location variables in particular were weakly associated with irrigation application rates (AF/AC). The precipitation variable took on a negative sign in district 72, from which it may be inferred that moister basins are less heavily irrigated. More significantly, irrigation rates were observed to closely parallel the seasonal length of irrigation (DAYS). Both of these diversion variables displayed a negative association with basin rank (MRANK) indicating that senior users tend to have much heavier irrigation application rates.

When the basin rank variable is considered more closely, however, a curious divergence is observed between district 70 and other districts. The negatively signed coefficient for district 72 may be interpreted as follows: senior appropriators practice higher irrigation application rates. This supports the popular view that the priority system encourages waste and inefficiency. The results in district 70, although not statistically significant, suggest just the opposite, i.e. that senior users tend to engage in more efficient water use practice, thus buttressing an opinion strongly endorsed by advocates of the appropriation doctrine. This interesting split led to more detailed investigation of priority and water use efficiency using individual diversion records.

Seniority and Water Use

Recall the hypotheses that senior appropriators divert more water, irrigate more land, irrigate for a longer period, and at more excessive rates of application than juniors. Mean basin ranks were computed from the water rights tabulation as a rough approximation of relative seniority among the 52 stream basins (figure 18). The map of mean basin ranks indicates relatively higher priority on the Colorado River main stem in all districts and in the Plateau Creek tributaries. The most junior priorities are concentrated, on the other hand, in the Roan Creek tributaries (district 70), Parachute Creek, the Salt Creeks in district 72, and in tributaries of district 45. Mean basin ranks were significantly associated with the seasonal length of water diversion revealing that more senior water users are able to irrigate longer. At present, however, differences in seniority have more effect on the allocation of water

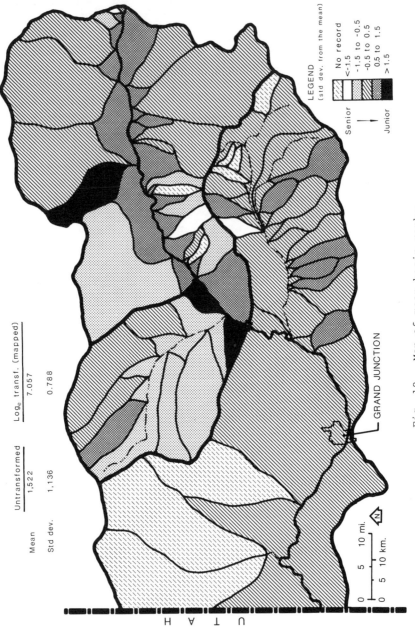

Fig. 18. Map of mean basin ranks

within stream basins than among them. Consequently, individual diversion records were used in place of basin averages to evaluate the effect of seniority (RANK) on diversion variables. It should be noted that several water rights are generally exercised in a single diversion structure. This makes it nearly impossible to sort out precisely which water rights are being employed at any given time.

When individual decree ranks were employed in the analysis, it became clear that seniority affects water diversion practice in a variety of ways, though not uniformly over the entire study area (Appendix, table 37). Of the five water diversion variables analyzed, only the length of the diversion variable (DAYS) gave unambiguous results for the study area as a whole.

Seniority clearly affects the length of the irrigation season. Interestingly, correlation coefficients were highly significant for districts 45 and 72 where scarcity and competition are most intense, but insignificant for districts 39 and 70. This is as expected because water commissioners implement the prior appropriation doctrine by closing the headgates of junior appropriators as water becomes scarce. Senior appropriators have also tended to locate at the most reliable stream locations, leaving riskier opportunities for those who followed.

One might expect that if the seasonal length of diversion is directly affected by relative seniority, then other water use variables such as diversion volume might follow closely, but the results are ambiguous. In district 39, where recent storage development improved local water supplies, diversion rates (CFS) were observed to vary indirectly with stream priority, as expected, but overall diversion volumes (AF) displayed the opposite relationship--junior appropriators use larger volumes of water. In district 45 on the other hand, juniors use significantly smaller volumes of water, illustrating again the importance of seniority in more water scarce locations. In no district was seniority significantly associated with the acreage under irrigation, and only in district 39 did diversion rates (CFS) appear to be affected by relative priorities. The scale of water use therefore does not appear to be a function of seniority.

Most puzzling were the findings for irrigation application rates. The hypothesis that senior water users are more wasteful and inefficient was supported only in district 72 where major Colorado River diversions account for the bulk of all water use. The null hypothesis was strongly supported, however, by the results in district 70, the Roan Creek drainage. Why would junior appropriators account for the most excessive rates of water use in this district (indeed in the study area, for district 70 had the highest average application rates among all four districts--see table 18)? The most plausible explanation lies in the relatively ineffective nature of water control in district 70. Debris laden spring runoff washes out water control structures more frequently than in other areas. It is possible that the most senior appropriators selected locations where water was less abundant, yet could be more easily diverted and controlled, leaving juniors to cope with the washouts and flash flood conditions of better endowed locations.

Seniority only exerts an important influence on actual water use patterns in certain areas--generally those in which water is most scarce and storage most limited. Even then the effects may vary or be outweighed by other factors as was observed in the case of both irrigation application rates and diversion rates. Only the seasonal length of diversion variable (DAYS) unambiguously fulfilled the research hypotheses. These findings bear directly on the consideration of alternatives for integrated water development. The search for waste, inefficiency, or available water should not be limited to the most senior water rights; it may exist throughout the water rights queue. Second, the water management needs of different areas were found to vary from an inadequate length of diversion and application rates in district 45, to inadequate water control in district 70, and excessive application rates in districts 70 and 72. Chapter seven takes up the issues of relative water use efficiency in greater detail.

Summary

Analysis of water diversion records provides a clearer picture of regional water supplies and water management problems than that afforded by water rights patterns alone. Administration of water diversions, for example, does not currently meet either the needs of local water users or the statutory responsibilities of the Colorado Division of Water Resources. Water measurement has necessarily been limited to the largest diversions and the most pressing conflicts, but as water demand increases and diversifies, the administrative resources of Water Division 5 will become even less sufficient to meet its growing responsibilities. More accurate estimation of actual water use will also become increasingly important for planning and conflict resolution.

Compilation of diversion records revealed that, even after allowing for incomplete records, annual water diversions range between only a quarter and a half of the total sum of decreed water rights. If correct, this suggests a "paper rights" problem on a scale that in other states, such as Texas, has led to general basin-wide adjudication proceedings. Colorado already has adequate administrative procedures to correct this problem but has not implemented them. The argument that abandonment lists and other administrative reforms would imperil Colorado's compact-apportioned share of the Colorado River would probably fail to hold up if a serious bid for reapportionment were made by the Lower Basin states. Paper rights are unlikely to fool such avid water seekers, whereas it is likely that these excess appropriations confuse and hamper water administration and exchange within the region.

Water diversion patterns varied greatly over the study area. Water appears to be most intensively used in district 45 where diversion seasons are short and application rates are low. Irrigators there naturally felt that the most pressing regional water management problem is a lack of reservoir storage water. In contrast, water use in the Roan Creek drainage (district 70) and in district 39 is characterized by heavy application rates and longer diversion seasons. For district 70, effective water control represents a chronic water problem. In districts 39 and 72, on the other hand, adequate reservoir storage and diversion facilities provide the most reliable water supplies in the region. In these last two districts, urbanization and industrialization present the most serious problems for effectively administering water diversions and improving water use efficiency.

Opportunities for integrated water development lie in the resolution of these specific problems and the trends which underlie them. To understand how the process of integrated water development might operate, it is necessary to evaluate these trends, as well as the formal mechanics of change, in greater detail.

CHAPTER VI

CHANGING PATTERNS OF WATER USE

Changes in water use represent a major theme in research on the Upper Colorado River Basin. Numerous studies have attempted to project changes in water use, water demand, and water quality. The more sophisticated studies make a range of assumptions regarding institutional mechanisms for water transfers, reuse, and water rights changes.[1] Rarely, however, have these studies investigated the actual processes of change in water use and ownership for insight into how, when, or where changes occur in water management.

In this chapter discussion focuses first on the institutional rules and procedures for making changes in water rights. Although there is considerable flexibility, these rules establish a body of precedents which all proposed changes must follow or reinterpret in some creative fashion. The role of flexibility and creativity becomes evident when these formal rules for water rights changes are contrasted with the record of actual water rights changes in the study area.

After the institutional framework for changes in water rights has been presented, changes in actual water use are considered. Recorded variations in the volume of water used, irrigated acreages, and irrigation application rates are described and then compared with annual variation in precipitation, farm product prices, and net farm income. Detailed evaluation of changes in water rights and water use may thus provide guidance for adjustments in water management practice that are particularly appropriate to local conditions in western Colorado.

---

1. John V. Krutilla and Constance Boris. *Water Rights and Energy Development on the Yellowstone River Basin* (Baltimore: Johns Hopkins University Press, 1979); and A. Bruce Bishop and Rangesan Narayanan, "Competition of Energy for Agricultural Water Use," *Journal of the Irrigation and Drainage Division, ASCE* 105 (1979): 317-335.

Changes in Water Rights

Colorado has historically allowed water rights changes, relying on a judicial forum for evaluating proposed water rights changes and foregoing the legislative proscriptions and administrative permitting that have been adopted in other appropriation States.[2] Water rights changes include a broad array of actions on either absolute or conditional rights (C.R.S. 1973, 37-92-103(5)).

1. Changes in the type, location, or timing of use
2. Changes in the means or point of diversion or storage
3. Changes from a fixed point of diversion or storage to alternate points, or *vice versa*
4. Changes from direct application to storage and subsequent application, or *vice versa*
5. Any combination of the above.

Related changes in water rights may also include abandonment, plans of augmentation, and conversion of conditional rights to absolute rights.

Rules Governing Water Rights Changes

Two fundamental legal rules--the "no injury" rule and the "historic beneficial use" rule--apply to water rights changes in Colorado. The no injury rule was adopted early in Colorado case law. "One who has acquired the right to divert the waters of a stream may change the point of diversion and place of use without losing his right of priority, where the rights of others are not injuriously affected."[3] Concepts of injury have evolved over the years from protests against enlargement of irrigation diversions to the following array of legal injuries:[4]

1. Increases in water consumption whereby more water is evapotranspired or removed from the basin, and less water returns to the stream for subsequent appropriation
2. Increases in stream conveyance losses that occur in effluent i.e. "losing," streams as a result of downstream transfers (this is recognized as injurious in several states but not Colorado)

---

2. For a comparison of State water laws, see Frank V. Trelease and Dellas W. Lee, "Priority and Progress: Case Studies in the Transfer of Water Rights," *Land and Water Law Review* 1 (1966): 1-75; and George A. Gould, "Conversion of Agricultural Water Rights to Industrial Use," preprint of July 1981 proceedings the Rocky Mountain Mineral Law Foundation conference in San Diego, CA.

3. *Sieker v. Frink*, 7 Colo 148, 2 P. 901 (1883). "The right to change either the point of diversion of a water right or its place of use is always subject to the limitation that such change shall not injure the rights of subsequent appropriators," *Farmers Highline,* 129 Colo. 575 (1954).

4. Gould, "Conversion of Agricultural Water Rights," passim.

3. Increases in the timing or length of the diversion season--either for irrigation or in conversion to some year-round use, e.g., domestic or municipal use
4. Changes in the location of diversions which affect return flow patterns (and subsequent appropriations)
5. Changes in the timing of return flows (which are temporarily stored in the soil profile)
6. Changes in water quality.

A widely adopted guideline for avoiding injury limits water rights changes to the historic consumptive use, though in some cases injury may be much less than total non-consumptive use while in other cases even transfers of consumptive use may cause injury. This leads to the eventual *reductio ad absurdum* that no transfers be allowed, thereby fully safeguarding against injury. The Colorado courts, on the other hand, have supported the view that injury can only be justly determined on a case by case basis and not by proscription. Colorado case law on water rights changes closely approximates the dictates of the Coase Theorem, that is, changes are never outright prohibited but are rather negotiated through stipulations on the amount to be transferred, precautions against injury, and compensation to the injured.[5]

The rule of historic beneficial use was derived originally to prevent sale of decreed water that had never been put to beneficial use, i.e., in which the acquisition of a water right had never been fully completed. Application of the benefical use rule is also site and situation specific. For irrigation uses it is comparable with the "duty of water" concept--that quantity of water needed for consumption by the crop, plus a reasonable amount of water for conveyance and operational losses. In Division 5, early adjudication set the duty of water at one cfs per fifty acres or .02 cfs/acre.[6] These decreed standards are not enforced, having given way in case law to the purposefully ambiguous definition of benefical use. Recently, however, the courts have laid emphasis on maximum beneficial use, stressing the growing importance of full utilization of the resource and the avoidance of waste. Waste and overappropriation have always been dimly viewed but only infrequently challenged as irrigators do not want to encourage bureaucratic regulation of water diversions.

---

5. Ronald H. Coase, "The Problem of Social Cost," *Journal of Law and Economics* 3 (1960): 1-44.

6. Interview with Lee Enewold, Division 5 Engineer in Glenwood Springs, CO; standards were set forth in the 1903 adjudication. Most water rights are exercised only when needed, but it should be noted that 0.02 cfs running continuously amounts to a total volume of over 7 acre-feet over the course of a 180 day growing season.

In actual water practice, numerous changes in the use of water rights occur from season to season and sometimes permanently without ever entering the courts for approval or record. Examples include shifts in crops or cropping patterns that result in different water consumption levels; extension or contraction of ditches and lands under irrigation; construction of farm ponds and wells; and changes in headgate locations. Technically, any such changes would require acquisition of a new right or perhaps modification of an old one. These changes occasionally present problems for field administration of priorities, but more frequently surface in attempts to sell or alter the right in some way at a later date.

Obviously, the unrecorded changes in use discussed above can greatly complicate the task of determining historic beneficial use. For example, how long must a particular crop or field be in cultivation to be used as the basis for computation? These issues generally lead to conflicting formulations of historic consumptive use that have to be resolved through negotiation and litigation. The advantages of the Colorado system lie prinicipally in its flexibility; its disadvantages lie in the transaction (e.g., legal) costs for water rights changes.

Procedures for water rights changes have been similarly flexible, but were made more efficient and sophisticated in the 1969 Water Right Determination and Administration Act. The 1969 Act requires that applications for changes in water rights follow a procedure in which: (1) an initial determination of the facts and law surrounding the change is made by a water referee, (2) public notice is issued of the proposed change, (3) objections are filed by interested parties, and (4) judgement is then made by the water court. In the absence of objection, the court frequently follows the referee's report. Strong objections may result in the case being re-referred for amended findings or stipulations on the proposed change. Stipulations may include a plan of augmentation, abandonment of part of the right, or installation and monitoring of water measuring devices. Often the right to future objection based on the discovery of unanticipated injuries is asserted. Appeal of water court decisions is made directly to the State Supreme Court, and final court action on the proposed change, if favorable, becomes part of the water rights tabulation.

The Record of Water Rights Changes in Western Colorado

The final record of water transfers, alternate points of diversion, and abandonments is documented in the water rights tabulation (table 14). The water rights tabulation does not record the conflicts, negotiations, and stipulations which underlie water rights changes. To remedy this, a review was made of Division 5 Water Court case files for all water rights changes listed in the tabulation as having occured between 1970 and 1977 (table 14). Contents of the ninety case files included referee's reports, objections filed, stipulations, and court judgements on proposed changes. If the case involved objections or stipulations, it was classed as a "contested" change, whereas if the referee's finding of "no injury" held up, it was classed as an "uncontested" change. The following types of water rights changes were evaluated:

1. Water Rights Transfers (which include changes in the point of diversion)
2. Alternate Points of Water Diversion
3. Water Rights Abandonment
4. Changes in Water Use.

Water rights transfers commonly occur in canal extension or consolidation projects, reservoir projects, and land transfers among irrigators, as evidenced in the relatively high number of transfers in district 39 (table 14). All districts show a concentration of water rights transfer activity on larger creeks, though several relatively small creeks also display considerable activity, as seen in Canyon Creek, Beaver Creek, and Mesa Creek. The overwhelming number of transfers involve a change in the point of water diversion *within* the same tributary basin, as evidenced in the net transfers row of table 14.[7]

Alternate points of diversion provide the irrigator with greater operational flexibility. They are sometimes sought in the expectation of future changes in use or to ensure future flexibility in the use of a right. In the past, they incurred less vigorous objections than did transfers. The geographical distribution of these actions parallels that of transfers though at far lower frequencies and with one curious exception in district 70 (table 14). During the rapid acquisition of water rights and the escalation of litigated conflicts in the Roan Creek drainage, a small number of water rights holders obtained numerous alternate

---

[7]. Net transfer data illustrate the small magnitude of extrabasin transfers. These figures probably give over-estimates of transfers due to poor cross-tabulation and double reporting of tranfers in Colorado water data bank records.

TABLE 14

SUMMARY OF WATER RIGHTS CHANGES BY DISTRICT

Variable	All Streams n = 52	District No.			
		39 n = 8	45 n = 13	70 n = 8	72 n = 23
Full Record[a]					
1. No. of Water Transfers[b]	378	191	87	33	67
2. Gross Volume of Transfers (cfs)[b]	6,412.75	632.15	113.00	419.18	5,248.42
3. Net Volume of Transfers (cfs)[c]	-22.32	-72.50	-9.75	+52.50	+7.43
4. No. of Alternate Points of Diversion	115	37	16	51	11
5. No. of Abandonments	18	0	13	1	4
Sample Record (1970-1977)[d]					
1. Change in Point of Diversion (uncontested)	33	15	5	6	7
2. Change in Point of Diversion (contested)	8	4	3	0	1
3. Alternate Point of Diversion (uncontested)	19	4	3	0	1
4. Alternate Point of Diversion (contested)	10	3	5	2	0
5. Change of Use (uncontested)	12	3	5	1	3
6. Change of Use (contested)	8	3	3	2	0
Total Uncontested Changes	64	22	17	9	16
Total Contested Changes	26	10	11	4	1
All Changes	90	32	28	13	17
% Contested	28.9	31.3	39.3	30.8	5.9

[a] From the 1978 Water Rights Tabulation, 1882-1978.

[b] Inclusion of all transfer citations makes this number and the gross volume of transfers roughly double their true values.

[c] Net transfers = transfers to-transfers from.

[d] Sample of all w-case files, roughly 1970-1977.

points of diversion to secure the future flexibility of their operations. This remains something of an anomaly that has not occurred in similar basins, such as Parachute Creek, Clear Creek, or Battlement Creek.

Although abandonments are rarely pursued, for the statutory and political reasons discussed above, several precedents exist for them in the study area. A flurry of abandonments occurred in district 45 in the adjudication of 1903, perhaps in early recognition of the poorer water supply in those tributaries. A more recent and dramatic abandonment was ordered for an unused conditional flow right to 100 cfs by the Orchard Mesa Irrigation District.[8] Technically the case proceeded as the cancellation of a conditional right, based on failure to exercise due diligence as opposed to actual abandonment of an absolute right. The significance of this action has not been lost on other water rights holders in the region. Calls for conservation, more accurate records of water use, enforcement of a duty of water, and elimination of "paper rights" are widely perceived as efforts to strip individuals of their property rights to water.[9] More frequent challenges of abandonment of absolute rights and of due diligence claims for conditional rights are expected in the entire Upper Basin system. Such proceedings will probably feature participation by transmountain diverters, public interest groups, and energy firms.[10]

Changes in the decreed use of a water right have naturally been sought in areas of urbanization and industrialization, although sometimes irrigators will seek to shift water from irrigation to stockwatering, domestic, or recreational uses. Changes in use have occured on a massive scale in the Front Range metropolitian corridor of Colorado and similar trends were projected for western Colorado. The low yield of water rights changed from irrigation to nonagricultural uses, coupled with the high frequency of legal objections to changes in use, has dampened interest in such changes. To avoid such obstacles in the future, new decrees are generally filed for a full range of agricultural and non-agricultural uses.

---

8. Division 5 court case no. W-168.

9. The legal underpinnings and widely divergent legal opinions on these matters will be taken up in greater detail in the following chapter on conservation practice.

10. Union Oil Company, for example, appears in water court as a frequent objector to water rights changes in order to protect its conditional and absolute rights on the river. Participation by advocacy groups, seemingly encouraged by the 1969 Act, has not been fully tested in the courts. See Musick and Cope, "An Introduction to Colorado Water Law," unpublished report (Boulder, CO, 1981).

Water rights changes occurred most frequently on the Colorado River, larger tributaries, and tributaries affected by prospective energy development. When water rights changes are mapped by stream basin, the Colorado River in district 45 displays the highest frequency of change (figure 19). District 39, which has the most imminent prospects for synfuels development, had the most uniformly high frequency of water rights change. Very low frequencies of change were recorded for smaller tributaries in each of the four water districts.

These findings were reinforced by the results of correlation between the number of water rights changes in individual basins and a set of stream basin, water rights, and water diversion variables (Appendix, table 38). The frequency of water rights changes tends to be greater in larger basins (AREA and YIELD), in basins with currently high rates of water diversion (CFS), and in basins with high rates of conditional water rights appropriation (CFLOW). Location variables, although sometimes locally strong, are often distorted by the importance of conditional water demand or basin yield. Water rights changes are not only more frequent in basins with high rates of conditional appropriation, such as Parachute Creek, they are also more controversial.

Objections to water rights changes and stipulated agreements stem from common legal principles and precedents, but have not yet yielded a clear pattern of results. Water rights changes sought by municipalities and energy firms, particularly "changes in use," invoke high levels of protest compared with similar changes sought by irrigators. The lowest percentage of contested cases in all categories was observed in district 72 (5.9 percent) perhaps due to its downstream location with respect to energy development. In each of the other districts, contested cases represented approximately a third of all cases and were concentrated in the basins most affected by energy development (figure 19 and table 14). Because stipulated agreements are specific with respect to site and situation, the following vignettes of five stipulated agreements are offered for illustration.

1. *Upstream transfer and change of use (W14)*--In 1970 three energy firms sought to transfer 11.11 cfs and 4.62 cfs from two rights on the Larkin Ditch on the Colorado River to a point 12 miles upstream at Parachute Creek. Objections were filed by irrigators, Chevron Shale, and the City and County of Denver. Three years later a stipulated agreement was reached among the applicants and objectors that conceded the following:

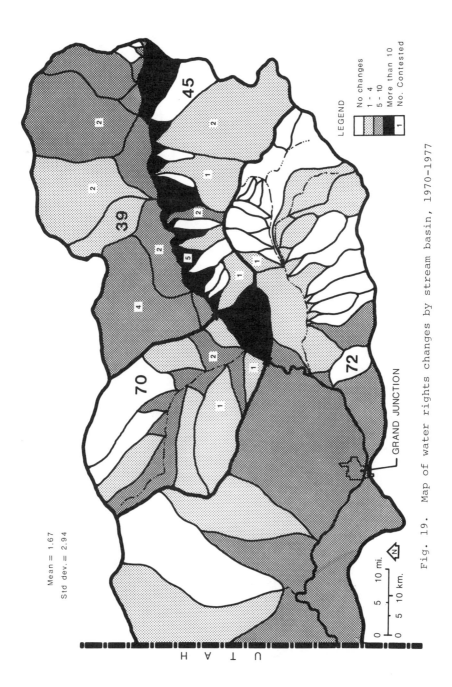

Fig. 19. Map of water rights changes by stream basin, 1970-1977

a) Transfer of only 2.5 cfs to the new location, not to exceed 800 acre-feet per year
b) Abandonment of 8.11 cfs to the stream
c) Maintenance of 0.5 cfs in the original location to avoid injury to other ditch users
d) Agreement not to seek transfer of Bluestone Ditch water rights
e) Nor to irrigate lands from which the water had been transferred

2. *Downstream transfer and change of use (W2206)*--In 1974 Union Oil Co. filed for water rights transfers from the Roaring Fork Drainage downstream to a pumping plant at the confluence of Parachute Creek with the Colorado River. Union's proposal also included changes in use, a plan of augmentation, and exchange of Parachute and Roaring Fork decrees. Objections were lodged by upstream irrigators and Denver. The final court decree stipulated that:
    a) Transfers in the Parachute Creek drainage occur only between April 15 and October 31, not to exceed a total consumptive use of 1395.9 acre-feet
    b) Water from the Roaring Fork not be used for irrigation in Parachute Creek and only be transferred when other supplies for oil shale operations are inadequate
    c) Transfers from the Roaring Fork occur only between April 15 and October 31, not to exceed 1152.6 acre-feet *less* carriage losses between the old and new points of diversion
    d) Annual notification of the Division Engineer as to whether agricultural rights will be transferred
    e) Installation of measuring devices
    f) Forfeit of the right to irrigate in any year that some portion of the right is transferred for energy use

3. *Alternate points of groundwater withdrawal (W2303)*--The Colony Oil Shale Project applied for alternate points of diversion for four wells in Parachute Creek (.444 cfs each). Objectors claimed potential injury from the proposed changes and obtained the following stipulations:
    a) Installation of measuring devices
    b) Maintenance of minimum surface stream flows
    c) Withdrawals not to exceed 800 acre-feet per year

4. *Changes in farm headgate locations*--Although changes in agricultural water use are less frequently challenged than

other types of change, conflicts are increasing in frequency (see W2044 on Beaver Creek, W1706 on Divide Creek, and W2665 on Elk Creek). In most cases these disputes were settled by leaving some portion of the water right in the ditch or by diversion scheduling agreements

5. *Transfers from tributaries to the main stem (W2127)*--In 1972 the Riverbend Development Corporation sought to transfer water rights on Canyon Creek to a point of diversion on the Colorado River for use in its subdivision. Denver objected strongly to the change in location as well as to proposals for water reuse, the argument being that main stem withdrawals could conflict with transbasin diversions whereas tributary uses could not. The final decree stipulated reductions in the volume of water to be transferred and measurement of consumptive use.

A general outline of water conflict may be inferred from contested changes in water rights (Appendix, table 39). Correlation of the number of contested changes with a selected set of stream basin, water rights, and water diversion variables reveals again the importance of basin yields (YIELD), diversion rates (CFS), and conditional appropriation (CFLOW). Contested water rights changes differ from the pattern for all water rights changes, however, in their strength of association with water rights seniority (MRANK). The negative correlation coefficient indicates that changes involving senior water rights are more vigorously challenged than those involving more junior water rights. Changes in senior rights pose a broader threat of injury to junior water users, particularly those changes which attempt to capitalize on "paper rights." Consequently, western Colorado finds itself in something of a double bind in which water rights are not efficiently distributed and in which redistribution can result in significant transaction costs or penalties for water rights holders.[11]

Analysis of Changes in Water Use

Two major periods of water rights activity have been identified--first, an early surge of appropriation spanning 1880 through 1920; and second, a mix of conditional appropriations, reservoir development, water rights changes, and well permits occurring from the mid-1950's to the present. In constrast, patterns of actual water use have, with a number of exceptions,

---

11. Penalties can result from "injury" to other rights as well as from cancellation of rights which fail to meet the criterion of historic beneficial use.

remained relatively stable over the past 80 years. Exceptions include the construction of Federal reclamation projects at Silt (1965), Grand Valley (1915), and Collbran (1964); the early spread and subsequent retreat of fruit cultivation along the Colorado River main stem; and suburbanization of irrigated farmland over the past decade. Short-term fluctuation in irrigated acreage and water diversions, on the other hand, may be observed on an annual basis.

Investigation of annual variation in water diversions and irrigated acreage was undertaken to identify recent trends in water use, and to evaluate explanatory variables for fluctuations in water use. The Division Engineer's annual reports from 1945 through 1979 provide rough estimates of total diversions, total irrigated acreage, and irrigation application rates by district (Appendix, table 40). Unfortunately, the measurement errors in these estimates almost surely pose a serious problem, although it is not possible to precisely determine the magnitude of error. In the first place, acreage figures are not measured by the Divsion Engineer's staff, but rather obtained from structure owners. Secondly, the annual reports make repeated reference to the poor quality or absence of measuring devices in the study area as well as to the perennial personnel constraint limiting the collection of a full record. These shortcomings notwithstanding, the Division Engineer's records represent the most accurate annual estimates of regional water use in western Colorado.

As a partial check on measurement error, the Division Engineer's annual diversion records were compared with the 1980 water commissioners' annual water diversion summaries. Mean values for district water use variables conform fairly well with 1980 diversion records evaluated in chapter five, except for mean irrigation application rates in district 72, which are higher than those in 1980, and mean irrigation application rates in district 70, which are lower than those of 1980. In district 72, the high mean irrigation application rate is in line with irrigation practice among senior appropriators in the Grand Valley area. The lower 1980 value may signal in part the start of the Grand Valley Salinity Control Program as well as the low rates of irrigation application observed in moister tributaries of the upper Plateau Creek drainage. In district 70, on the other hand, the flashy nature of tributary streamflows (as evidenced by the high coefficients of variation for both water diversion rates and irrigation application rates) perhaps accounts for the discrepancy between 1980 and early periods of record. As is pointed out below, irrigation application rates in district 70 have also been increasing in recent years.

Trends in Water Use

Significant trends during the 1960-1978 period were observed for several water use variables (Appendix, table 41). Irrigated acreage, for example, has declined significantly in district 39, perhaps due to urbanization, water rights transfers, and reservoir construction. Irrigated acreage has increased in district 70, on the other hand, perhaps due to installation of more effective water control structures and perhaps due to an effort to demonstrate greater beneficial use of decreed water rights (undertaken in anticipation of future transfers of those rights). Diversion volumes have increased in districts 70 and 72 and in the study as a whole, as have irrigation application rates. Thus, more water is being diverted per acre of irrigated land, and consequently, irrigation efficiency would appear to be declining in the study area. In district 45, by contrast, the only area where mean irrigation application rates do not greatly exceed crop requirements, irrigation application rates have declined in recent years. Significant autocorrelation coefficients for diversion volumes in district 70 and irrigated acreage and application rates in district 39 reinforce the observation of trends in these variables.

Annual Variations in Water Use

Variation in agricultural water use may result from a wide range of causes, including climatic variation, shifts in product prices, or fluctuations in net farm revenue--any combination of which can lead to adjustments in water demand.[12] Which of these factors are most closely related to annual variations in water use? Are the hypothesized relationships widespread or localized in extent? To answer these questions, the Division Engineer's annual water use records (volume diverted--AF; acreage irrigated--AC; and irrigation application rate--AF/AC) were correlated with the following explanatory variables:

1. *Product Prices* (indexed to 1967 and deflated by the purchasing power of the dollar)
    a) Colorado Livestock Prices (WMC)
    b) National Livestock Prices (WMUS)
    c) Colorado Hay and Forage Prices (WHAY)
    d) Colorado Fruit Prices (WFC)

---

12. Technological change (e.g. conversion to sprinkler or drip irrigation) and changes in cropping patterns could also significantly affect water use, but, as will be pointed out in chapter seven, these have not occurred in the study area on a large scale.

    e) Index of All Colorado Farm Product Prices (WALLC)
2. *Net Farm Revenue* (WREV--indexed to 1967 and deflated by the purchasing power of the dollar)
3. *Annual Precipitation*
    a) At Grand Junction, Colorado (JUNC)
    b) At Rifle, Colorado (RIFLE).

It was hypothesized that product price and farm income variables would be positively correlated with water use--as prices or revenue increase, water use and irrigated acrage should also increase. As precipitation increases, on the other hand, water use ought to decline. It should be noted that price and income variables displayed strong positive autocorrelation, but no trend, between 1960 and 1978. Prices and income rose sharply in the early 1970's and have declined just as sharply in the past several years. Precipitation variables showed neither trend nor serial correlation.

In spite of the absence of clear trends in the explanatory variables that might have matched those observed among water use variables, two sets of correlation coefficients were produced. In the first set, annual water use values were correlated with the same year's prices, revenue, and precipitation values (Appendix, table 42).[13] In the second set of correlation coefficients, explanatory variables were lagged one year to allow for a period of adjustment between causal events and their effects (Appendix, table 43).

In the first set of coefficients, product prices and net farm revenue had the most important and interesting patterns of association with water use variables (Appendix, table 42). Livestock prices (WMC) and overall prices (WALLC) seemed to exert a stimulating effect on diversion volumes and irrigation application rates in the study area as a whole, whereas hay prices had stronger local effects.[14] Hay prices were positively correlated with irrigation application rates and diversion volumes as expected. Correlation with irrigated acreage, however, was positive for district 72, but negative with districts 39 and 45. The negative

---

13. Precipitation records for Rifle and Grand Junction came from *U.S. Climatological Data,* annual summaries for the period 1960-1979. Price indexes were obtained from volumes of *Colorado Agricultural Statistics* and are indexed to a 1967 base year. The deflators--purchasing power of the dollar and value of the dollar--are indexed to the same base year. Net farm revenue data were obtained from the same sources.

14. For more detailed modelling of these relationships, see Gaylord V. Skogerboe et al., *Potential Effects of Irrigation Practices on Crop Yields in Grand Valley,* EPA Report No. 600/2-79/149 (Ada, OK: U.S. Environmental Protection Agency, 1979); Robert E. Howitt, W.D. Watson, and R.M. Adams, "A Reevaluation of Price Elasticities for Irrigation Water," *Water Resources Research* 16 (1980): 623-628; and Ronald D. Lacewell and Gary D. Condra, *The Effect of Changing Input and Product Prices on the Demand for Irrigation Water in Texas,* Technical Report no. 75 (College Station, TX: Texas A & M University, Texas Water Resources Institute, 1976).

coefficient for district 39 is probably spurious due to the negative effects of urbanization on irrigated acreage in that district. To correctly interpret the results in district 45 we would have to know by what extent declines in irrigated acreage were offset by higher irrigation application rates, i.e., what had happened to the tonnage of hay harvested per acre. Increases in net farm revenue also appeared to stimulate increased water use, particularly diversion volumes in districts 70, 72, and the study area as a whole as well as with irrigated acreage in district 70. Interestingly, very few of the other price or precipitation variables had any significant association with the water use variables. Precipitation records displayed only a weak negative association with irrigated acreage in district 72, possibly explained as a reduction of irrigated pasture and marginal lands in wet years.

To evaluate the hypothesis that last year's precipitation and prices affect water use decisions more than this year's, lagged correlation models were developed (Appendix, table 43). Lagged precipitation values were positively associated with water diversion and irrigation application rates in districts 45, 72 and for the study area as a whole, but they were negatively associated with irrigated acreage indicating that increases in precipitation lead to more intensive irrigation of smaller parcels of land, whereas lower precipitation is accompanied by more extensive irrigation, probably of pasture rather than cropland. Most important, however, was the broader significance of State and national prices for livestock products, possibly reflecting the effects of higher livestock prices and smaller herd sizes on the demand for feed.[15] Hay prices declined in importance. Lagged revenue variables did not offer substantially different correlation results from unlagged values, except for the more pervasive effect on irrigated acreage both locally and in the study area as a whole. To summarize, some variables such as livestock prices, precipitation, and net farm revenue seem to have more significance in the following year's water use decisions, while other variables such as hay prices have immediate significance.

Summary

This study has touched on issues of change in water use at several points: first, in presenting the evolution of the "law of the river" and the appropriation doctrine in Colorado (chapter 3);

---

15. Cattle-feed relationships have complicated most efforts to estimate the marginal value product of western irrigation water. Bishop and Narayanan, "Competition of Energy;" and Michael D. Frank and Bruce R. Beattie, *The Economic Value of Irrigation Water in the Western United States: An Application of Ridge Regression,* Report no. 99 (College Station, TX: Texas A & M University, Texas Water Resources Institute, 1979).

second, in the analysis of early water rights appropriation patterns and more recent conditional decrees (chapter 4); and third, in evaluating the effects of seniority on water diversion practices (chapter 5). In this chapter a more direct series of investigations focused on changes in water rights and water use.

Rules and procedures for water rights changes in Colorado are widely regarded as among the more flexible, equitable, and efficient applications of the prior appropriation doctrine. The record of water rights changes in western Colorado raises some doubts, however, about their operation in practice. The market for senior water rights can hardly be described as "organized," let alone "efficient." Knowledge of the availability of water rights is incomplete; assessment of the value of water rights is often grossly inaccurate, particularly in advance of transfers or changes in use; the risks of having to abandon (or constrain) some part of the right are high but erratically applied (depending upon the participants involved); and transaction costs are excessive.

Proposals to create a "water market" for either water rights or water loans have not proceeded very far, although they hold considerable potential for alleviating at least the information and communication problems mentioned above. As was the case with both water rights records and water diversion records, the public interest is not being adequately served by inadequate recording of shifts in water diversion, irrigated acreage, or water use. Accurate records could provide at least a partial basis for a water market; they would greatly facilitate implementation of the "no injury" and "historic beneficial use" rules; and they would permit rational planning for trends and fluctuations in the State's water resource needs.

The emphasis on *change* in water use patterns in this chapter is perhaps misleading. Water use and irrigation practice have been remarkably stable in the study area. To be sure, construction of Federal reclamation projects has improved agricultural water supplies in the study area, the 1969 Water Rights Determination and Administration Act has established the groundwork for more effective water administration, and the latest synfuels boom has led to a significant amount of agricultural land conversion. Still, water use has not changed in many important ways. Few trends were observed in water use variables. Irrigated acreage is declining in areas of urbanization. Irrigation application rates are increasing in all areas except district 45--the area where water is most needed. These higher rates are perplexing. Are irrigators using more water; are more irrigators now able to divert the traditionally high rates of water due to improved water control; or are appropriators seeking a legal record of excessive decrees by diverting more water and flooding more land in hope of someday selling those rights to speculators? Whatever the answers to these questions may be, they probably do not mean that the methods, practices, or exchange relations among water users have changed in any fundamental way.

Annual fluctuations in water use, however, are common. Short-term shifts in hay prices and net farm revenue appear to have immediate effects on water use, while livestock prices and precipitation have slightly lagged effects. Interestingly, as precipitation increases, so does the intensity of irrigation--fewer acres are irrigated, but at higher rates of application. When viewed in the context of a trend toward higher rates of irrigation in the study area as a whole, this suggests that water use is part of a general trend toward agricultural intensification in the region. If so, it would seem that more effective means of water control and water conservation could serve the same goal while serving also the needs of emerging water users.

Finally, it should be emphasized that there exists no single type of water "surplus" or "shortage" in western Colorado but rather a variety of each. The classic shortage, where water is applied for short periods in insufficient quantities over large areas, occurs in district 45. In the Roan Creek drainage (district 70), on the other hand, water is available in abundant yet erratic and uncontrolled quantities. Surplus seems more common in districts 39 and 72 from the aggregate statistics, but local irrigators in those areas will describe how one farmer floods his sagebrush pasture while the next farmer's hay crop browns out.

Water conservation may be defined at least in part as the efficient balancing of surplus and scarcity among interdependent water uses. The remaining work takes the findings of chapters 4, 5, and 6 a step further by identifying linkages between local water surpluses and shortages in the study area and giving emphasis to solutions that coordinate historical and emerging patterns of water demand.

CHAPTER VII

WATER CONSERVATION PRACTICE IN WESTERN COLORADO

Water conservation has a broad range of meanings and motives. Conservation implies an improvement in the efficiency of water use. In irrigation this involves using water more judiciously over the growing season in order to meet crop requirements while minimizing water waste, soil erosion, salinization, and other adverse impacts of water use. For some in western Colorado, conservation thus means more reservoirs, while for others, whose notion of conservation also includes the preservation of natural landscapes, it means avoiding new dam construction through ditch lining, irrigation scheduling, water reuse, and so on. Moreover, as Soil Conservation Service personnel have indicated, irrigators often adopt conservation practices for non-conservation reasons, such as for improving cropping patterns or reducing labor requirements and other farm costs. Consequently, simple estimates of improved water use efficiency do not provide sufficient information on the causes or broader effects of water conservation activity.

In this chapter an assessment is made of water conservation and its prospects in western Colorado. The potential role of water conservation for alleviating regional water supply problems has been raised on several occasions, only to fall prey to institutional constraints, claims of economic infeasibility, or simple prejudice. Chapter three, for example, raised the issue of strategic behavior in Upper Basin water development. Recent regional water assessments cite higher relative costs for implementing conservation practices than for groundwater development, perfection of conditional decrees, or purchase of Federal reservoir water.[1] The Colorado Department of Natural Resouces study estimates that a 130,000 acre-feet reduction

---

1. Colorado Department of Natural Resources, *13(a) Assessment* (1979); and U.S. Office of Technology Assessment, *An Assessment of Oil Shale Technologies* (Washington, D.C., 1980).

in incidental water losses would cost almost 700 million dollars, or approximately $5,000 per acre-foot. Simple market transfers of water from irrigation to energy uses, on the other hand, were estimated to cost between $1,000 and $2,500 per acre-foot--far above the current costs of purchasing water from Reudi Reservoir (table 11).[2] Most of these studies acknowledge that other benefits, such as improved water quality, recreation, and farm productivity are subsumed in the higher costs; it should also be emphasized that their cost estimates represent average costs over a very large area and not the marginal cost of conservation improvements or water transfers in local areas.[3] Finally, rules and insitutions governing water rights changes pose an equally powerful set of constraints on water conservation.

The research approach adopted here parallels that of earlier chapters. First, institutional rules and procedures bearing on the adoption of water conservation measures are charted out. A survey of existing patterns of conservation is then made with detailed attention given to recent adoption of conservation measures in the Grand Valley Salinity Control Program. The salinity control program is administered in conjunction with the Agricultural Conservation Program (ACP). Salinity control represents the largest and most extensively documented water conservation program which has taken place in the study area to date. Grand Valley irrigators' expressed conservation needs are compared with actual patterns of conservation investment. The conclusion emphasizes the importance of water organizations in conservation and, in doing so, seeks to critically extend the conventional property rights paradigm for water resources management.

Conservation in Colorado Water Law

Historically, as conflict over scarce water supplies has escalated, water law has evolved to deal with more sophisticated issues of conservation. In an early complaint against Colorado's record of over-appropriation and of allowing changes in the location and type of water use, Elwood Mead stated that,

> In every instance investigated the real purpose of water rights changes has been to make money out of excess appropriations. The parties who have acquired surplus rights are unable to use the water themselves, and seek to sell to someone who can (p. 174) . . .

---

2. The actual delivered cost of Reudi Reservoir is said to be approximately $200.00 per acre-foot due to stream losses and other costs. Roland Fisher quoted in "Why Will Water Cost So Much?" *Weekly Newspaper* (Glenwood Springs, CO), February 1981.

3. The marginal value product of irrigation water has been estimated at $14.54/af in Garfield County and $17.80/af in Mesa County by A. Bruce Bishop and Rangesan Naryanan, "Competition of Energy for Agricultural Water Use," *Journal of the Irrigation and Drainage Division, ASCE* 105 (1979): 317-24.

> As the adjudications were held remote from the ditches and the land affected, and the judge or referee did not visit the lands and ditches to see for himself whether the situation was as it had been described, appropriators were encouraged to make extravagant claims (p. 149) . . .
>
> What they prefer to do and what they are attempting to do is to loan the water which they do not need, and by doing so put parties having late priorities in the place of the holders of earlier rights who would otherwise receive it (p. 177).4

Chapter six outlined the two rules--"no injury" and "historic beneficial use"--that govern water rights changes. These rules serve to prevent the abuses Mead describes,5 but they also encourage over-irrigation and inhibit transfer of surplus water.

Although an appropriator acquires no right to waste water, the definition and regulation of excess water use has generally been limited to nuisance cases. Even then the Colorado Supreme Court has refrained from quantifying the basic dimensions of wasteful use or unreasonable irrigation practice. In 1947, the Court held:

> . . . we do not think the time has yet been reached in the State, when the owners of such enterprises can be held to such a high degree of diligence in their construction as to be compelled to prevent them from seeping at all . . . as the result of such a rule would mean, in most cases, that the costs or means to prevent the seepage would be far in excess of the value of the properties so damaged.6

Later, in *Fellhauer v. People,* the Court's opinion on water use efficiency was changing:

> As administration of water approaches its second century, the curtain is opening upon the new drama of *maximum utilization* and how constitutionally that doctrine can be integrated into a law of *vested rights*. We have known for a long time that the doctrine was lurking in the backstage shadows as a result of the accepted, though often violated, principle that the right to water does not give the right to waste it.7

This view was also put forward as a supporting argument in 1979 in *A-B Cattle v. United States,* the "silty water case," to counter an irrigation company claim that it had a right to the historic silt load in its diversion which had reduced seepage from its unlined canals.

> In using its leaky ditches the Bessemer Co. has not attempted to make maximum utilization of the water . . . the plaintiffs do not have the right to use the silt content to help seal leaky ditches.8

The difficulty of integrating the concepts of maximum utilization and vested rights arises from the fact that increased water use efficiency by definition involves a change in the historic pattern of use and often an increase in the net volume of water

---

4. Elwood Mead, *Irrigation Institutions* (New York: Macmillan Co., 1903).

5. See *Enlarged Southside Irrigation Ditch Co. v. John's Flood Ditch Co.* 116 Colo. 580, 183 P.2d 552 (1947).

6. *Middlecamp v. Bessemer Irrigation Co.,* 46 Colo. 114, 103 P. 280 (1909).

7. *Fellhauer v. People,* 167 Colo. 320, 447 P.2d 986 (1968).

8. *A-B Cattle Co. v. United States,* 196 Colo. 539, 589 P.2d 57 (1979).

consumed as well. In practice, on-farm conservation improvements that enable greater consumptive use by crops, e.g. an additional hay irrigation, generally go unchallenged. The State only monitors diversions from a stream, and neighbors are unlikely to interfere with each other's activities unless gross damage or nuisance occurs. Let the same individual attempt to transfer or sell the water he has saved, rather than use it himself, and objections will be filed throughout the basin.

In a similar fashion, the Colorado courts have held that "salvaged" water--water which has historically gone to "waste" but which has through some action become available for use--is still subject to the priority system and is not free from the call of the river.[9] In the *Shelton Farms* case, the Court denied decrees to applicants who had cut down phreatophytes and then claimed first right to the water that had previously been lost through transpiration.[10] Only "developed" waters, those that were not formerly part of the basin hydrologic system (e.g. transbasin diversions), are free from the call of the river and subject to reuse.[11] The answer, then, to the question of who reaps the benefits of water saved through conservation would appear as follows: in practice, the farmer who saves water may alter his production patterns to take some advantage of the savings *de facto* but in a strict legal sense, the water rights queue should benefit from savings made by senior water users, not proportionately but in order of priority. The conserving farmer may usually only sell or transfer a very small portion of the water conserved.

What incentives exist then to invest in conservation improvements (figure 20)? Assume A is the most senior right on the stream. In the simplest case, the most senior rights after A, particularly those upstream, have the strongest incentive to pay for conservation measures that will reduce A's diversion requirement. Very junior rights benefit only if the entire queue reduces its diversions, e.g., through massive public investment in conservation. Downstream juniors may even lose by A's conservation. If the transaction is conducted as a water rights transfer or change in the location of use (say from A to D), then even the most junior appropriator (say B) can object to the sale by claiming injury due

---

9. Wells A. Hutchins, *Water Rights in the Nineteen Western States,* vol. II, U.S. Department of Agriculture Miscellaneous Publications, no. 1206 (Washington, D.C.: Government Printing Office, 1974), pp. 565, 567.

10. *Southeastern Colorado Water Conservancy District v. Shelton Farms, Inc.,* 187 Colo. 181, 529 P.2d 1321 (1975).

11. *City of Denver v. Fulton Irrigating Ditch Co.,* 179 Colo. 47, 506 P.2d 144 (1972).

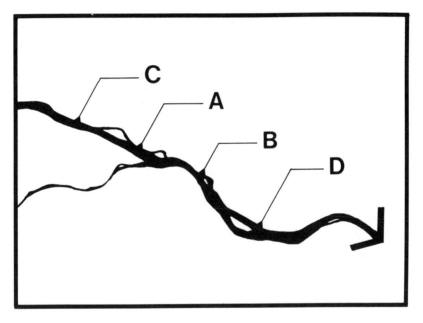

Fig. 20. Diagram of relative user locations in a stream basin

to shifts in diversion and return flow patterns. No objection can be made, however, if the transaction involves abandonment of a portion of a right, as junior appropriators have no guarantee to continued diversion by a senior user.

Three important principles emerge that will guide the empirical investigation of conservation practice and potential in western Colorado. First, a water right is compound in nature as it allows diversion or storage of one quantity, and transfer of some smaller portion of that quantity. Although only the "beneficially used" portion is transferable, in many cases conservation problems, such as salinity, water-logging, and erosion, arise from the exercise of excess diversion rights. Thus, the merger of vested rights and maximum utilization will depend on a more critical assessment of the nature and limits of diversion rights. Second, any water conservation practice involves an implicit transfer of water, but that transfer may or may not be subject to strict administration. Is the example of the individual farmer who lines his canal and benefits therefrom (contra *Shelton Farms*) extensible to larger entities, such as ditch companies, irrigation districts, or Federal reclamation projects; or should conservation by those entities result in a reevaluation of their diversion right?[12] Third, as the Court recognized in the *Shelton Farms* case, conservation has many dimensions and often conflicting motivations. Clearing of phreatophytes to reduce transpiration, for example, may accelerate stream bank erosion, destroy important habitat, increase turbidity, and reduce the scenic quality of an area.

Existing Conservation Practice in Western Colorado

Given the diverse character of motivations and activities inherent in conservation, how might the status of conservation activity in western Colorado be described? Chapter five provided a survey of actual water diversions, irrigation application rates, and irrigated acreage. The relative absence of water reuse and rental arrangements, reservoir storage, canal linings, and sophisticated ditch structures or sprinkler systems in the study area was highlighted. The Division Engineer's annual reports on water use suggest that little has changed by way of improvements in irrigation efficiency over the past several decades (figure 21).

---

12. "There is no question that one who merely clears out a channel, lines it with concrete, or otherwise hastens the flow of water is not entitled to a decree therefore." (citations omitted). *Southeastern Colorado Water Conservancy District v. Shelton Farms, Inc.*, 187 Colo. 181, 529 P.2d 1321 (1974).

133

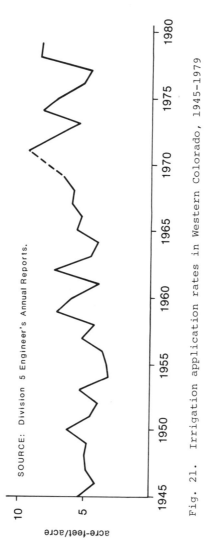

Fig. 21. Irrigation application rates in Western Colorado, 1945-1979

Chapters three and four discussed the State's role in water development and administration; the initiation of a water rights tabulation and and abandonment lists; and procedures for water rights changes. Water Division 5 has historically had the loosest approach to administration of water diversions and conflicts in Colorado. Even insistence by the Division Engineer on the installation of headgates met with resisitance from water users.[13] Most public and private water managers in Division 5 seem to support the current practice of having the State and water supply organizations "keep their noses out of water use and ownership" at the farm level. The Division Engineers put it this way, "there's no water *management* here; that's not our job."[14] In contrast, Water Division 6, which includes the White River basin, has surveyed local water uses and developed a computer program to estimate evapotranspiration.[15] The justification offered for these activities emphasized their value for conflict resolution both in the fields and in the courts.

The major exception to this traditional, casual water practice in Division 5 has occurred in the Grand Valley Salinity Control Project. Massive investments in canal lining and on-farm improvements have been made to reduce saline return flows. The conservation research and investments made in Grand Valley offer an opportunity to evaluate proposals for integrated development involving water exchange and conservation.

Current Conservation Practice in Grand Valley, Colorado

The Grand Valley area contributes between 600,000 and 700,000 tons of salt annually to the Colorado River, derived principally through leaching by surface irrigation return flows, and to a lesser extent, tailwater runoff.[16] Water withdrawals for the Grand Valley Project in 1978 were roughly 322,770 acre-feet to irrigate 31,830 acres, or 10.14 acre-feet/acre.[17] Of the net supply of 10.14 acre-feet/acre, 4.99 acre-feet were lost in conveyance or spills and 5.26 acre-feet were delivered to the farm. This represents the highest average delivery rate of all Federal reclamation projects in

---

13. Division 5 Engineer's Annual Reports (Denver, CO, 1971).

14. Interview with Lee Enewold, Division 5 Water Engineer (Glenwood Springs, 1981), on the distinction between water mangement and administration.

15. Telephone interview with Kent Holt, Colorado Water Division 6, Steamboat Springs, 1981.

16. U.S. Soil Conservation Service, *On-Farm Program for Salinity Control* (Denver, CO, 1977).

17. U.S. Bureau of Reclamation, *Water and Land Resource Accomplishments, Statistical Appendix I* (Washington, D.C., 1978), p. 271.

the Upper Colorado River Basin. Estimates of potential evapotranspiration in Grand Valley have ranged from 1115 mm (3.7 feet) to 1309 mm (4.3 feet) for the period of May through October.[18] When monthly values of evapotranspiration were multiplied by crop coefficients, seasonal crop evapotranspiration for a well-watered crop was estimated to be 650 mm (2.13 feet).[19] Thus, on-farm irrigation efficiencies are approximately 50 percent and overall water use efficiency only 30 percent--low even by regional standards.[20]

An intensive series of salinity research projects has focused on alternative irrigation practices for increasing water use efficiencies and reducing saline return flows. Conservation practices studied have included: canal lining, lateral ditch lining, irrigation scheduling, drainage, improved agronomic and cropping practices, desalting of return flows, pipelines, gated pipe, and sprinkler delivery systems. The applicability and feasibility of such measures have been frequently assessed from an engineering economics standpoint, providing useful material for this investigation and making further review unnecessary.[21] Responsibility for implementing the salinity control program lies jointly with the Bureau of Reclamation (canal lining) and the Soil

---

18. Gaylord V. Skogerboe et al., *Potential Effects of Irrigation Practices on Crop Yields in Grand Valley* (Ada, OK: Robert S. Kerr Environmental Research Laboratory, 1979), pp. 67-68.

19. Ibid., p. 71. Crop coefficients were adjusted from M.E. Jensen, *Consumptive Use of Water and Irrigation Water Requirements* (New York: American Society of Civil Engineers, 1973). See also U.S. Soil Conservation Service, *Irrigation Water Requirements,* Technical Release no. 21 (1970); and idem, *Crop Consumptive Irrigation Requirements and Irrigation Efficiency Coefficients for the United States* (1976).

20. On-farm irrigation efficiency is defined as the ratio of crop water requirement to the amount delivered to the farm, where the crop water requirement is given as:

Crop water requirement = ET(crop) + LR - P(e) - W(soil)

Where: ET(crop) = crop evapotranspiration

LR = leaching requirement

P(e) = effective precipitation

W(soil) = soil moisture

Overall efficiency is defined as the ratio of the crop water requirement to total project water supply. See M.G. Bos and J. Nugteren, *On Irrigation Efficiencies,* Publication no. 19 (Wageningen: International Institute for Land Reclamation and Improvement, 1974).

21. Gaylord V. Skogerboe et al., *Evaluation of Drainage for Salinity Control in Grand Valley* (Washington, D.C.: U.S. Environmental Protection Agency, 1974); idem, *Evaluation of Irrigation Scheduling for Salinity Control in Grand Valley* (Washington, D.C.: U.S. Environmental Protection Agency, 1972); idem, *Potential Effects of Irrigation Practices on Crop Yields in Grand Valley* (Ada, OK: U.S. Environmental Protection Agency, 1979); Robert G. Evans et al., *Evaluation of Irrigation Methods for Salinity Control Technology in Grand Valley* (Ada, OK: U.S. Environmental Protection Agency, 1978); and Wynn R. Walker, et al., *"Best Management Practices" for Salinity Control in Grand Valley* (Ada, OK: U.S.

Conservation Service (on-farm irrigation improvements).[22]

Previous research and administrative reports have highlighted the importance of irrigators' perceptions and behavior as well as the institutional arrangements for implementing the conservation program. A comprehensive Soil Conservation Service survey of current water use practices and perceived conservation needs in 1975 permitted inquiry into the conditions under which conservation practices have been adopted in the Grand Valley area.

The 1975 SCS Survey collected information on more than 8,000 fields in the Grand Valley area for use in planning the on-farm improvement program. The survey provided basic reconnaissance information, field contact with irrigators, and an inventory of on-farm improvement needs. Irrigators were asked about their current irrigation practice and their anticipated adoption of conservation practices, assuming "generous" cost-sharing and technical assistance.[23]

The interpretation of current conservation patterns is based on joint frequency distributions of conservation practice types (in place prior to the salinity control program) by canal service areas. A sample was taken of 384 fields to evaluate the distribution of current conservation activity by canal service area.[24] Canal service areas operate under different water distribution rules and were consequently felt to offer greater interpretive power than other measures of relative location in the study area (figure 22).

The twenty canals and canal segments sampled are operated by six major irrigation organizations (table 15).
Four of the organizations, marked by an asterisk, receive Federal reclamation project water which is appurtenant to specific tracts of land, regardless of land use. The remaining two organizations are incorporated mutuals whose water shares may be traded or sold as land uses or water demands shift, but only to other shareholders within the company's service area.[25]

Environmental Protection Agency, 1978).

22. Memorandum of Agreement between the USBR and USSCS (1979).

23. SCS personnel indicated that the implied level of assistance was necessarily vague.

24. Sample size was based on random sampling for the largest canal service area, approximately 40 percent of the study area, with a precision of ±5 percent. Because the sampling fraction is less than 5 percent, no finite population correction was required. The final sample population ranged from 359 to 367 fields depending on individual conservation practices.

25. Chapter 8 will differentiate among organizational arrangements in greater detail. The principal point to be made here is that the Grand Valley area is irrigated by incorporated mutuals, irrigation districts, and water users associations.

TABLE 15

WATER SUPPLY ORGANIZATIONS AND MAJOR CANALS
IN THE GRAND VALLEY AREA

	Organization	Canals Operated
1.	Grand Valley Water Users Assoc.*	Government Highline Canal
2.	Palisade Irrigation District*	Price Ditch
3.	Mesa County Irrigation District*	Stub Ditch
4.	Orchard Mesa Irrigation District**	Orchard Mesa Power Canal Orchard Mesa #1 (high) and #2 (low)
5.	Grand Valley Irrigation Company	Grand Valley Canal Grand Valley Main Line and Highline Kiefer Extension Mesa County Ditch Independent Ranchman's Ditch
6.	Redlands Power and Water Co.	Redlands Power Canal Redlands #1, #2, and #3

*Participating in the Grand Valley Project, Garfield Gravity Division.

**Participating in the Grand Valley Project, Orchard Mesa Division.

Land use varies more from east to west than among individual canals, with the easternmost area supporting orchard and vegetable crops on very small fields; a transition zone on either side of Grand Junction; and the western two-thirds of the Grand Valley area supporting corn, barley, and forage crops on larger acreages (Appendix, table 44). The sample consists of fields, not farms, that range in size up to 58 acres but average only 8.8 acres (Appendix, table 45). Over 70 percent of the fields sampled are smaller than ten acres in size. Corrugation and furrow irrigation methods and unlined delivery ditches dominate in all service areas with infrequent practice of wild flooding, border, and sprinkler methods. The existing distributions of five conservation practices are considered in subsequent paragraphs: flow measuring devices, gated pipe delivery, ditch lining, pipeline delivery, and sprinkler systems.

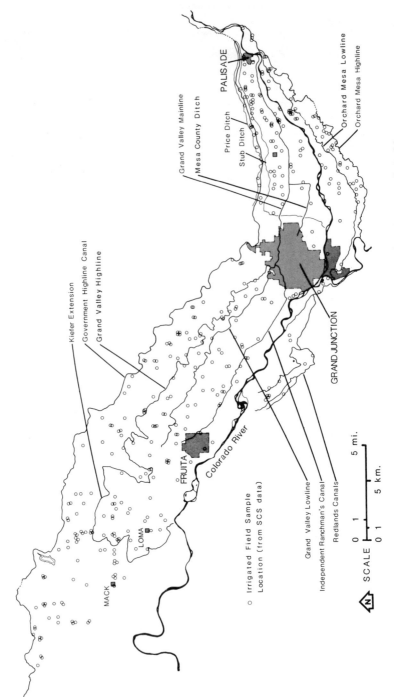

Fig. 22. Map of major canals and sample survey sites in the Grand Valley area

Flow Measurement

Flow measurement is a prerequisite for scientific estimation of irrigation efficiency and for irrigation scheduling. A summary engineering evaluation of Grand Valley irrigation stated, "Proper water management requires a strong emphasis toward on-farm water control structures, especially flow measurement devices."[26] At the time of the survey, measuring devices had been installed for only 13 percent of the surveyed fields, though this figure may understate the proportion of farms using measuring devices due to the small size of individual fields and multiple field holdings of most farmers. Contiguous fields may be served in many cases by a single flume or flow metering device. In each case only one measuring device was recorded per field, regardless of field size. Tabulation of mean frequencies, given field size and canal number, indicated that such devices were most frequently employed for fields of 11 to 29 acres in size, i.e., one to two standard deviations above mean field size. Grand Valley Project water users under the Government Highline and Orchard Mesa canals accounted for over 90 percent of the measuring devices recorded, contrasting sharply with the absence of such devices among mutual ditch company irrigators (Appendix, table 46).

Piped Water Delivery

Piped water delivery systems minimize water losses through evapotranspiration and seepage while allowing improved cropping practices and reduced maintenance. Gated pipe and pipeline delivery systems were employed even less frequently than measuring devices as of 1975, presumably due to high capital costs. Gated pipe accounted for only 2.7 percent of all on-farm conveyance methods by length and pipelines were only slightly greater at 4.7 percent. The installation of pipeline delivery systems was again skewed toward fields receiving Federal project water, particularly the Orchard Mesa Division and Palisade Irrigation District, though not to the extent evidenced for measuring devices (Appendix, table 47). The Redlands Canals also had a relatively high proportion of pipeline systems, uncharacteristically so in light of their very low rates of adoption and interest in other conservation practices.

---

26. Wynn R. Walker et al., *"Best Management Practices" for Salinity Control in Grand Valley* (Ada, OK: U.S. Environmental Protection Agency, 1978).

Sprinkler Systems

Pressurized irrigation systems provide the greatest precision for irrigation scheduling and the most uniform rates of application in areas of topographic irregularity.[27] Two types of systems have had application in the Grand Valley area--first, sideroll and other moveable systems for forage crop irrigation on larger fields; and second, trickle irrigation for orchard crops in the Palisade area. Three sample fields (less than one percent) had sprinkler systems, one each under the Government Highline, Grand Valley Mainline, and Orchard Mesa Highline; no trickle irrigation systems were recorded in the sample.

Ditch Lining

Lining of earth ditches and major laterals constituted the most popular and reportedly most cost-effective water conservation practice in the study area. Laterals from the main canals to farm headgates carry an average flow of five cfs; they have an estimated total length of over 600 kilometers, and they contribute approximately 32 percent of all subsurface return flows and salt loading from the Grand Valley area.[28] The sample inventoried 53.4 miles of ditches, of which approximately 10.3 miles (or 19 percent) were lined. The distribution of these improvements, however, diverges from the pattern seen thus far (Appendix, table 48). Fields under the Grand Valley Irrigation Company canals, particularly the Mainline, Highline, and Kiefer Extension, had slightly greater than average frequencies of canal lining. Only the Government Highline and Price Ditch among the Grand Valley Project Canals had comparable frequencies, while Orchard Mesa and Redlands were far lower than average. Over 80 percent of the lined ditches occured on fields less than ten acres in size suggesting that field size did not pose the same constraints reported for adoption of sprinkler methods.

---

27. Some SCS personnel suggest that sprinkler systems are not necessarily more efficient than well designed border or furrow systems and are not applicable to all cropping patterns due to drift in larger systems and limited spray areas in trickle or drip systems (Interview in Denver regional SCS Office, 1981). However, Walker concludes that "achievable application efficiencies for sprinkler irrigation are 85 to 95 percent depending on the level of management of the system," ibid., p. 32.

28. Robert G. Evans et al., *Evaluation of Irrigation Methods for Salinity Control in Grand Valley* (Ada, OK: Robert S. Kerr Environmental Research Laboratory, 1978), p. 47.

Summary

The five conservation practices inventoried represent a conservation baseline that was shown to vary among canal service areas. The frequency of existing conservation practices in the Grand Valley area as a whole varied from less than 5 percent (gated pipe and pipelines) to almost 20 percent (lined ditches). In almost every case high frequencies of occurance were recorded for the Government Highline Canal operated by the Grand Valley Water Users Association. The Orchard Mesa Division of the Grand Valley Project ranked similarly high for all but ditch linings. Fields under the mutual ditch companies, on the other hand, ranked relatively low on the adoption of conservation practices with the notable exception of ditch linings. Of the canals operated by these mutual companies, fields under the Kiefer Extension and western segments of the Grand Valley Canal had the highest recorded frequency of adoption while the Redlands Canals, Independent Ranchman's Canal, and Mesa County Ditch had relatively low frequencies of adoption. Any interpretation of these results would have to include the effects of: distribution shortages and conflicts, as on Orchard Mesa; the seniority of Grand Valley Irrigation Company water rights relative to those of the Grand Valley Project; the appurtenancy requirement of Federal reclamation water; and the inferior canal positions of irrigators in western Grand Valley--all of which would tend to force greater investment in conservation practices.

Perceived Conservation Needs

Chapter three noted that whereas water conservation programs seek reductions in erosion and river salinity, adoption of conservation practices is based on a broader set of objectives related to improved water delivery and reduced farm costs. The Soil Conservation Service survey of improvement needs reflects this divergence in objectives. Irrigators were asked to estimate their conservation needs, assuming favorable cost-sharing assistance. Not all of the improvements inventoried, such as land leveling and field drainage, would result in major salinity reductions, yet irrigators may have responded in strategic fashion to gain support for these practices or to accomodate their water control objectives to the salinity cost-sharing program.[29]

---

29. Wynn R. Walker et al., *"Best Management Practices" for Salinity Control in Grand Valley* (Ada, OK: U.S. Environmental Protection Agency, 1978). Strategic behavior has frequently been postulated in surveys of willingness to pay for public investment but rarely proven in empirical investigations.

In 1977 the Soil Conservation Service compiled the survey results and roughly estimated their sensitivity to government funding levels (table 16). Based on this survey and additional planning information, the SCS proposed an on-farm program, the most recent cost estimate of which came to $38,088,000 (table 17). Our concern with perceived conservation needs follows directly from the discussion of current conservation practice. How do the frequency and distribution of interest in ditch lining, pipelines, and land leveling compare with current practices? How do perceived conservation needs vary among canal service areas? How do these patterns compare with the proposed on-farm salinity program? The following conservation practices are considered: measurement devices, pipeline delivery, ditch lining, land leveling, and field drainage.

Flow Measurement

It will be remembered that only 13.1 percent of the sampled fields had flow measurement devices in 1975. When questioned about the desire for such devices, this proportion rose to 39.3 percent of the sampled fields (Appendix, table 49). In absolute terms the greatest interest in flow measurement occurred in fields under the Government Highline Canal, but when adjusted for the number of observations in each service area, similarly high frequencies were observed for most canals. Only the Independent Ranchman's Ditch maintains a consistently low frequency of interest in flow measurement devices. This survey of expressed needs represents a much broader pattern of interest than that observed in the survey of existing conservation practice.

Piped Water Delivery

The frequency of interest in pipeline delivery systems increased from 4.7 percent of the sampled fields that currently rely on such systems to 12.6 percent that are said to need them. This moderate level of increase may signify high perceived capital costs, traditional reliance on gravity flow ditch delivery, and the lower cost-effectiveness of pipelines calculated by some engineers.

> Use of high-head PVC pipe (SDR 81 or greater) or concrete pipe is not a cost-effective alternative to ditch-lining or low-head PVC and should be discouraged. Attendant problems with the use of low-head pipe can be overcome by giving particular attention to design and rigorous installation specifications.30

---

30. Wynn R. Walker et al., *"Best Management Practices" for Salinity Control in Grand Valley* (Ada, OK: U.S. Environmental Protection Agency, 1978), p. 31.

TABLE 16

PROJECTED LEVELS OF LAND TREATMENT ON IRRIGATED
LAND APPLIED THROUGH A TEN-YEAR CONTINUATION
OF REGULAR PROGRAM ACTIVITY

Type of Practice	Units	Projected Quantity to Be Installed		
		Level 1	Level 2	Level 3
Applied by Individuals				
Ditch Lining	ft.	458,000	572,000	754,000
Irrigation Pipeline	ft.	122,000	152,000	218,000
Irrigation Structures	no.	200	250	400
Land Leveling	ac.	4,800	6,300	8,800
Land Smoothing	ac.	3,200	3,800	3,300
Open Drain	ft.	10,700	10,700	10,700
Closed Drain	ft.	105,000	154,000	187,000
Drip Irrigation Systems	ac.	140	160	350
Sprinkler Irr. Systems	ac.	180	200	430
Border Irr. Systems	ac.	130	150	200
Applied by Groups				
Ditch Lining	ft.	212,000	296,000	336,000
Irrigation Pipeline	ft.	24,300	32,000	50,800
Irrigation Structures	no.	160	200	450
Closed Drain	ft.	1,500	4,500	7,500

SOURCE: U.S. Soil Conservation Service, *On-Farm Program for Salinity Control* (Denver, 1977).

NOTE: Level one assumes that funding for cost-sharing assistance will be 25% less than the average for 1972-1977. Level two assumes cost-sharing equal to the current average, and level three assumes that assistance will be 25% greater than the current average.

TABLE 17

ESTIMATED COST OF CONSERVATION IMPROVEMENTS

Item	Unit	Estimated Quantity	One-time Installation Cost
Ditch Lining	ft.	2,882,000	$13,118,000
Ditch Structures	ea.	3,500	862,000
Pipelines:			
10" diameter	ft.	373,000	1,365,000
10"-15" diameter	ft.	248,000	1,460,000
15" diameter	ft.	3,200	39,000
Structures	ea.	1,800	268,000
Land Leveling:			
400 yd^3/ac.	ac.	4,100	737,000
400-800 yd^3/ac.	ac.	11,300	2,329,000
800 yd^3/ac.	ac.	1,500	587,000
Subsurface Drains	ft.	286,000	1,895,000
Measuring Devices	ea.	2,600	490,000
Drip Irrigation	ac.	300	381,000
Sprinkler Irrigation	ac.	500	699,000
Off-Farm Laterals:			
Pipeline:			
10"-15" diameter	ft.	680,000	4,504,000
15" diameter	ft.	250,000	5,358,000
Lined Open Ditch	ft.	75,000	1,038,000
Installation Cost - Irrigated Land			35,130,000
Non-Irrigated Land (Private-owned)			2,958,000
Total Installation Cost			$38,088,000

SOURCE: U.S. Department of Agriculture, Soil Conservation Service, *Supplement No. 1: On-Farm Program for Salinity Control* (Denver, 1980), p. 60, "alternative four."

A rather diffuse areal pattern of interest in pipelines emerged among private canal service areas and Federal project lands (Appendix, table 50). As in the case of measuring devices, the Independent Ranchman's Ditch recorded the lowest level of interest.

The total budgeted length of pipeline improvements relative to other conveyance improvements (21 percent of total conveyance length in the SCS budget cost estimate) paralleled the sampling results of perceived irrigation needs. This indicates that irrigators' expressed needs for conveyance improvements were translated fairly closely into SCS policy. The question may be raised again, however, concerning the extent to which irrigators' responses were conditioned by survey conditions and previous SCS funding arrangements.

Ditch Lining

Ditch lining drew the highest levels of interest among irrigators, reflecting a continuing trend toward upgrading traditional practice in water delivery. The proportion of fields needing such improvements averaged 53.6 percent among all canal service areas (up from an existing 18.5 percent) with the highest frequencies recorded for Orchard Mesa canals, the Grand Valley Canal, and Garfield Gravity Division Canals (Appendix, table 51). Characteristically low frequencies were observed for the Mesa County Ditch, the Independent Ranchman's Ditch, and Redlands canals. Favorable cost-effectiveness and demonstrated field performance of ditch lining probably led to its prominence in the farm needs survey and subsequent SCS on-farm program budget. Investigation of actual conservation investments made in 1980, however, will reveal a striking divergence from the emphasis on ditch lining.

Land Leveling

In spite of reservations expressed in some recent research, land leveling was second only to ditch lining in the inventory of improvement needs. Fully 43.7 percent of the fields sampled and 41.1 percent of the sampled acreage were believed to require land leveling. The object of land leveling lies in improving the uniformity of furrow irrigation applications. Above-mentioned reservations argue that low soil infiltration rates and relatively steep slopes make cut-back or pump-back irrigation methods more cost-effective than extensive land leveling.[31] Interest in land leveling was more evenly spread among canal service areas than was the case for most other improvements (Appendix, table 52). Although

---

31. Ibid., p. 36.

the Mesa County and Independent Ranchman's canal areas recorded no
interest in land leveling, the Redlands canals broke with that
pattern and followed the relatively high overall pattern of
interest. This high level of interest among irrigators, however,
was not proportionately matched in the SCS budget proposal in which
only 14 percent of all irrigated acreage is scheduled for land
leveling--less than one-third of the expressed demand for these
improvements.

Field Drainage

Drainage improvements were regarded as tangential to most
salinity control objectives. Their benefits lie more in improved
on-farm water management and reduced problems with field
waterlogging than reduction of saline return flows. Only 82 acres
on 22 fields were perceived to need drainage improvements. Most of
these needs lay in the western half of Grand Valley, especially
under Government Highline, Grand Valley mainline, and Kiefer
extension canals. Given this small response (less than 3 percent of
the sampled land area and less than 6 percent of the sampled
fields), SCS programming for 286,000 feet of subsurface drains seems
exaggerated and would have to based on some more compelling planning
information.

Summary

In summary, the inventory of perceived conservation needs
provides several important points of comparison with both existing
conservation practice and planning proposals. Ditch lining, flow
measurement, and land leveling stood out among the conservation
alternatives considered, representing an interest in upgrading
traditional practice. These improvements seem favored over more
substantial shifts in irrigation technology such as to sprinkler and
pipeline technologies. Planning proposals by the SCS closely follow
irrigators' perceived needs in most cases, though drainage proposals
appear over-represented and land leveling proposals
under-represented.

Whereas the presence of existing conservation practices
appeared concentrated on Federal project service areas, the level of
interest in the most important conservation practices--ditch lining,
pipelines, flow measurement, and land leveling--was more broadly
diffused among public and private entities. Conspicuous only by
their absence of interest in conservation improvements were several
privately operated canal service areas under the Redlands Canal, the
Independent Ranchman's Ditch, and the Mesa County Ditch. The most

plausible explanation for these low relative frequencies, aside from their small service areas (and resulting low sample representation), emerges from the SCS survey's data on "future land use" (Appendix, table 53). These three canals had the highest frequencies of expected suburbanization, and while pressures may be equally great in other canal service areas, such as the Price Ditch or Orchard Mesa Canals, the higher perceived rate of suburbanization would naturally have a muting effect on interest in agricultural water conservation. The fact that these areas are located in privately operated irrigation districts where water rights shares are separately negotiable from land transactions would further dampen interest in water conservation investments tied to specific tracts of land.

Rarely does the opportunity arise to compare survey data with actual behavior at a later date. In this case, the data collected on recent conservation expenditures in the Grand Valley area permit inquiry into the institutional, economic, and administrative contexts of conservation activity.

Actual Patterns of Conservation Investment in Western Colorado

We have said that the interpretation of water conservation activity becomes problematic as a result of uneven correspondence between motives and action. Does a reduction in water use due to changes in crop type or the cultivation of additional lands, for example, qualify as a conservation investment? When do private shifts in water use practice constitute a public interest in conservation? The historical record of Federal, State, and local assistance programs aimed at irrigation management bear evidence to the slippery nature of these questions. The Agricultural Conservation Program (ACP) has particular importance for this study--first, in explicitly confining its mission to conservation improvements; and second, in serving as the vehicle for implementation of the SCS on-farm salinity control program.

The Agricultural Conservation Program has its origins in the Soil Conservation and Domestic Allotment Act which established the Soil Conservation Service and authorized funding to assist the voluntary actions of local soil conservation districts and state committees.[32] Evolving fitfully and unevenly over a period of forty years, the ACP was given a clearer statement of purpose and eligibility requirements in the Food and Agriculture Act of 1977.[33]

---

32. Soil Conservation and Domestic Allotment Act of 1935 (49 Stat. 1148).
33. Food and Agriculture act of 1977 (91 Stat. 913).

... eligibility for assistance under the ACP should be based on the existence of a conservation or environmental problem which reduces the productive capacity of land and water resources or causes degradation of the environment. In determining the level of assistance, the Secretary is to consider the extent of the conservation or environmental benefits accruing to society; the cost of measures or practices; the degree to which appropriate practices would be applied in the absence of assistance; and the extent to which producers benefit from other conservation and environmental protection programs...... No assistance is to be offered for carrying out measures or practices that are primarily production-oriented or that have little or no conservation or pollution abatement benefits.[34]

Conservation Investment in Grand Valley

Water conservation and water quality have become increasingly important components of the Agricultural Conservation Program, as illustrated in table 18. Water conservation assistance, for example, represents over 90 percent of all ACP expenditures in Garfield County since 1979. The on-farm salinity control program in Grand Valley was funded as a special project of the ACP through the 1979 Agricultural Appropriations Act.[35] The salinity control project varies from ordinary ACP funding in the following ways:

1. The cost-sharing limit was increased from $3,500 per farm to $10,000 for each farm participating in a group project (pooling agreement)
2. Participants must install automated or semi-automated systems to be eligible for assistance
3. Cost-sharing levels, ordinarily ranging from 50 to 60 percent for irrigation improvements, were increased to as much as 90 percent of total construction costs for high priority needs (table 19).[36]

Previous discussion of customary conservation practice and expressed improvement needs provides a foundation for the interpretation of recent ACP and salinity control project expenditures on pipelines, ditch lining, land leveling, and other irrigation improvements.

Initiation of the Grand Valley salinity control project in 1979 increased ACP funding in Mesa County by nearly an order of magnitude (figure 23). A survey was made of all 97 cost-sharing assistance projects paid out by the Grand Junction ASCS office. Eleven of the 97 were individual on-farm projects, and 59 were on-farm pooling agreements; 27 were regular ACP pooling agreements; a total of 181 farms were involved. The seventy on-farm projects are considered in detail below.

---

34. U.S. Department of Agriculture, *National Summary Evaluation of the Agricultural Conservation Program, Phase I* (Washington, D.C., 1980) p. 4-5.

35. 1979 Agricultural Appropriations Act (92 Stat. 1073).

36. "Grand Valley Salinity Control Project," memo to participants (Agricultural Stabilization and Conservation Service, 1981).

TABLE 18

PERCENTAGE OF ACP PAYMENTS IN THE UNITED STATES
BY PRACTICE CATEGORY, 1940-79

Practice Type	Year				
	1940	1950	1960	1970	1979
Soil Loss	85.0	85.3	64.3	57.6	59.3
Water Conservation	8.7	9.8	18.1	21.7	20.4
Water Quality	.7	4.6	13.1	15.5	18.4
Forestry and Wildlife	1.5	.3	4.5	5.0	1.9

SOURCE: Agricultural Stabilization and Conservation Service, *National Summary Evaluation of the Agricultural Conservation Program, Phase 1* (Washington, D.C., 1980), p. 8.

NOTE: Figures do not necessarily add to 100 percent because some of the practices funded were not applicable to the categories selected.

A map of these projects reveals a remarkable concentration of activity, first, under the Orchard Mesa High Canal; second, in a band approximately five miles west of Grand Junction; and third, between Clifton and Palisade (figure 24). SCS officials providing technical assistance stated that their activity was diffused evenly throughout Grand Valley. Funding for project implementation has been concentrated, however, in part due to an informal agency desire to evaluate the effects of extensive improvements.

The areal distribution of total conservation investment and cost-sharing assistance revealed no apparent variation among types of canal organizations (Appendix, table 54). Public and private canal service areas in western Grand Valley had comparable levels of investment; these were nearly double the investment made in eastern Grand Valley. Interestingly, Orchard Mesa accounted for over 30 percent of both total investments and cost-sharing assistance. This parallels the high level of expressed improvement needs on Orchard Mesa. Several questions emerge for future consideration: to what extent does the distribution of cost-sharing assistance reflect actual patterns of interest and willingness to pay?; to what extent administrative bias?; and to what extent does the high participation of "downcanal" irrigators north of the river and of conflict-ridden

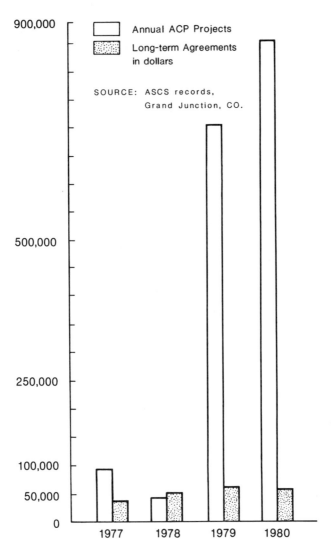

Fig. 23. Annual ACP expenditures in Mesa County, 1977-1980

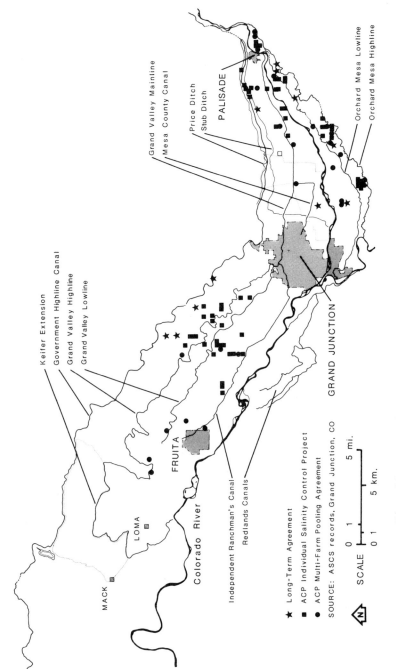

Fig. 24. Map of ACP projects in Grand Valley, 1980

irrigators on Orchard Mesa evidence the inferior water supply positions of those canal service areas?

Curiously, almost no projects are located in or very near either of the two salinity control demonstration areas. Not surprisingly, only one ACP project was located in the three canal service areas with disproportionately low levels of existing conservation practice and expressed improvement needs (Redlands Canals, Independent Ranchman's Ditch, and the Mesa County Ditch).

A dramatic reversal occured in the types of conservation practices selected. Whereas ditch lining and land leveling were most frequently cited in the needs survey, those practices represented relatively small fractions of actual expenditures (Appendix, table 55). Adoption of gated pipe nearly equalled that of ditch lining, and pipeline expenditures exceeded those of ditch lining by over 500 percent. Possible explanations for this shift include strong SCS emphasis on pipeline delivery systems and the higher level of federal cost-sharing for such systems (table 19). As illustrated in the Appendix, table 56, the distribution of different types of improvements among canal service areas closely paralleled the overall pattern of investment. This suggests a relatively uniform mix of conservation technologies adopted, in spite of differences in expressed preferences.

Conservation Investment Outside Grand Valley

The salinity control program represents a massive conservation investment when compared with regular ACP assistance in the rest of the study area. In 1980 the Mesa County district funded two small individual projects and one large pipeline project outside of Grand Valley; a total of 35 farms were involved and $17,598 cost-shared. In Garfield County, fourteen small projects were funded at 70 percent cost-sharing with a ceiling of $3,500, and one large pipeline project was assisted that involved 53 farmers. Expenditures in 1980 doubled those in 1979 and were five times as great as those in 1978 (figure 25). Among the small projects, pipelines, corrugated metal pipe, and ditch structures were the most frequent improvments adopted. A map of these projects reveals that none was located in district 70; most were located along the Colorado River in districts 39 and 45, with several projects scattered in the middle reaches of tributaries in the latter district (figure 26). Several SCS officials and water commissioners indicated that often it was the wealthier and younger farmers who participated, but that the low assistance ceiling discouraged many.

TABLE 19

MAXIMUM FEDERAL COST-SHARING ASSISTANCE BY WATER CONSERVATION PRACTICE, GRAND VALLEY SALINITY CONTROL PROGRAM, 1981

1. 90 percent of the cost:
   a. Pipe laterals including necessary valves, gates, trash kickers, sediment basins, and filters
   b. On farm automated or semi-automated underground pipelines including necessary valves, gates, trash kickers, filters, and flow measuring devices
   c. Automated or semi-automated ported concrete ditch lining including necessary fixtures
2. 75 percent of the cost of the following if they are components of a semi-automated or automated system
   a. Gated pipe
   b. Side role sprinklers
   c. Solid set sprinklers
   d. Drip or trickle irrigation systems
3. 75 percent of the cost for land leveling
4. 75 percent of the cost for installation of tile necessary to protect or make the practice or project function properly
5. 50 percent of the cost of installing non-automated concrete-lined ditches including laterals

SOURCE: Agricultural Stabilization and Conservation Service, memorandum (Grand Junction, Colorado, 1981).

The high proportion of pipeline projects along the Colorado River raises the question of whether this is associated with pumped delivery of Colorado River water in those locations or rather the need to improve delivery of water in downstream tributary locations.

Summary

Water conservation activity in western Colorado has been focused for the most part on the issue of river salinity. More sweeping proposals for conservation have encountered criticism for high costs, strategic constraints, and institutional barriers limiting their applicability. Although in some circiumstances junior appropriators and the public have incentives to finance conservation by senior appropriators, the no-injury and historic beneficial use rules for water rights changes tend to severely limit

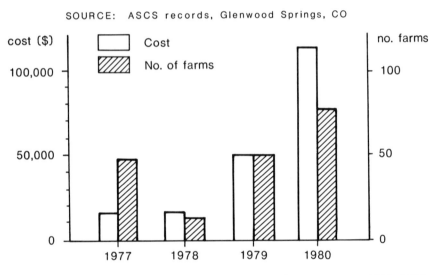

Fig. 25. Annual ACP expenditures in Garfield County, 1977-1980

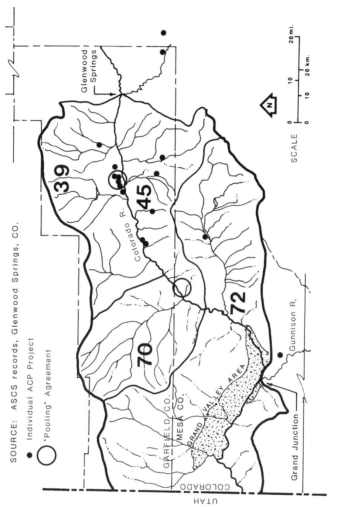

Fig. 26. Map of ACP projects in Garfield County, 1980

those circumstances. It was observed that junior water users, "downcanal" irrigators, and water-short areas of the Grand Valley engaged more frequently in water conservation activities. This supports the hypothesis that scarcity exists even in areas of excessively high average irrigation rates and senior water rights. It also illustrates that significant conservation investments will be made in high usage areas, given favorable cost-sharing levels. A national evaluation of the ACP program revealed that conservation investments in such high usage areas are the most cost-effective for reducing the volume of water diversions (table 20).[37]

TABLE 20

BENEFITS AND COSTS OF ACP WATER CONSERVATION ASSISTANCE, 1975-1979

Pre-assistance rate of water use	Average annual water savings	Percent of total annual water saving	Cost per acre-foot conserved (dollars)
(Feet/Acre/Year)	(Feet/Acre/Year)		
0-1	0.05	0.3	–
1-2	0.20	3.1	63.21
2-3	0.29	7.5	21.93
3-4	0.36	13.5	21.89
4-5	0.70	10.2	11.33
5-6	1.00	14.9	–
6-7	1.28	12.5	8.89
7-8	1.62	4.2	4.38
8	6.43	33.8	1.06
Total	0.76	100.0	9.88

SOURCE: Agricultural Stabilization and Conservation Service, *National Summary Evaluation of the Agricultural Conservation Program, Phase 1* (Washington, D.C., 1980), p. 38.

Proposals to transfer any portion of a diversion right in excess of that which is beneficially consumed, however, have met with fierce opposition from Mead's time onwards. Further clarification is needed as to the relationship between consumptive use and beneficial use. What effect would changes in the definition of "consumptive use" or "waste" have on the application of the

---

[37]. U.S. Department of Agriculture, *National Summary Evaluation of the Agricultural Conservation Program* (Washington, D.C., 1980).

beneficial use rule? The following critique on the common distinction between consumptive and non-consumptive uses, formulated over twenty years ago, goes part way toward clarfying the issues:

> Any use of water is always and necessarily consumptive with respect to other uses that are competitive with it; a given use can be non-consumptive only to the extent that potentially complementary other uses exist. Along a river, therefore, upstream irrigation may be non-consumptive insofar as return flow contributes to downstream uses. But a downstream irrigator's use will be entirely consumptive, irrespective of return flows, if there are no users farther downstream.38

As an extension or rather amendment of this view: any use is fully consumptive at the location and time of its use; it is fully consumptive with respect to all upstream junior users at all times; it is partially consumptive with respect to all downstream juniors who may enjoy use of the water (i.e., in return flows) at some later time; and it is non-consumptive only for senior water users, regardless of their location or time of use. Given the importance of instream and estuarine flows, there is no such thing as Hirshleifer and associates' final downstream user, after whom water is only wasted.

The principal policy issues surrounding the adjustment of diversion rights concern the interplay of upstream and downstream juniors. The former are "robbed" by excessive diversions, the latter "injured" by any reduction in diversions, and both cheated by improvements in water use efficiency without corresponding adjustment in diversion rights. In the Grand Valley case, where there are no major downstream uses in Colorado, irrigators are being paid by the nation to reduce return flows. The question remains as to whether or how they should be paid by upstream (e.g., transmountain) users to also reduce their historic levels of diversion.

Analysis of the Agricultural Conservation Program illustrated the influence of administrative behavior and organizational arrangements on expressed conservation needs and actual expenditures. Although the effects of SCS policies and procedures were not separable from Grand Valley sample results, areal concentration and technological bias in cost-sharing assistance were clearly indicated. The broader perspective of conservation activity throughout the study area revealed that larger irrigation organizations, for example, on the Bluestone, Grand River, and Grand Valley canals, garnered the major portion of ACP assistance. Because these organizations have developed internal rules and procedures for water allocation and exchange that differ in

---

38. Jack Hirshleifer et al., *Water Supply: Economics, Technology and Policy* (Chicago: University of Chicago Press, 1960), p. 67.

important ways from those of state administrative traditions, it can be argued that the conventional emphasis on property rights issues in water resources studies should be complemented by greater concern for the organizational fabric of water management.

CHAPTER VIII

THE ORGANIZATIONAL FABRIC OF WATER MANAGEMENT

Water institutions evolve in adaptive and transformative ways with respect to their environmental and economic contexts. Early surveyors of the American West recognized the limitations of existing policies and institutions for land settlement on the frontier.[1] The appropriation doctrine for water rights allocation is often cited as an adaptive response to scarcity and variability in water supplies. The priority rule provided tenure security for immovable capital investments in water diversion and conveyance facilities. Rejection of both the riparian doctrine and water rights transfer prohibitions provided tenure flexibility to meet changing economic and technological circumstances.

As recent evidence of change, transfer rules are being more rigorously enforced than in the past in order to avert legal injury; and inefficient water use practices receive progressively less sympathy from the courts as competition for water increases. The argument that scarcity promotes greater activity in property rights definition and enforcement is widely invoked to explain historic administrative deficiencies in western Colorado, but in fact the basic rules have changed only slowly over time.[2] Whereas this body of *rules* for water allocation has evolved in a relatively cautious manner, leading many to exaggerate the strength of custom in water law, the organizational fabric of water use reveals a record of experimentation and structural alterations that affect the practical

---

1. John W. Powell, *Lands of the Arid Region of the United States* (Washington, D.C.: Government Printing Office, 1879).

2. The scarcity explanation for property rights evolution has obvious intuitive appeal, see Walter P. Webb, *The Great Plains* (New York: Ginn & Co., 1931), and more recently Terry L. Anderson and P.V. Hill, "The Evolution of Property Rights: A Study of the American West," *Journal of Law and Economics* 18 (1975): 163-179. But qualifications are raised in Lawrence C. Becker, *Property Rights: Philosophic Foundations* (Boston: Routledge Kegan Paul, 1977) on p. 4 that devolve from claims to liberty in the acquisition of property. See also Richard Schlatter, *Private Property, the History of an Idea* (New Brunswick, NJ: Rutgers University Press, 1951).

operation of allocative rules.³

In this chapter organizational trends and arrangements which could ameliorate water managment problems in the study area are assessed. The discussion first reviews explanatory models for organizational evolution in western water development and then proceeds to evaluate the record of organizational change in western Colorado. Local differences in water organization provide a point of entry for reappraising the concept of integrated development and its potential application to specific water management problems in the study area. Four classes of organizational arrangements which could facilitate water conservation and exchange are then considered: water conservation on "company creeks," successive reuse agreements, multipurpose water development organizations, and, finally, organizational solutions to the "conserved water" problem.

Evolution of Water Organizations

The history of water development in the West records numerous ventures that failed or gave way to more successful arrangements. Interpretation of this record has frequently focused on the adaptive limitations of different organizational types. Most accounts give passing acknowledgement to the cooperative nature of early Hispanic community ditches, Mormon irrigation communities, and experimental agrarian communities such as the Union Colony at Greeley, Colorado. Formal property rights definition became important in areas of more heterogeneous settlement, but the principles of cooperative water development persisted in "ditch companies" and "mutuals" to provide a sustainable form of organization for irrigation development.

In contrast, Elwood Mead noted the recurrent failure of absentee speculative canal ventures during the 1890's, and he offered a more balanced explanation of their failure than the common complaint against speculative capital investment. Mead cited the panic of 1893, severe winters, overly optimistic engineering, conflict with public land laws, and the slow rate of return to capital as contributory to the failure of these ventures. He also recognized that, "Many of the largest and costliest canals had been built with English and Scotch capital . . . From 1870 to 1890 ditch building outran settlement. From 1890 to the present [1903], the West has been chiefly engaged in putting these canals to use."⁴

---

3. Here we are concerned primarily with water use organizations, i.e., contractual entities, and less with the process of rule-making or the behavior of executive entitites, though these may affect the powers and structure of different types of water organization. See Irving K. Fox, "Institutions for Water Management in a Changing World," *Natural Resources Journal* 16 (1976): 743-58.

4. Mead, *Irrigation Institutions,* pp. 344, 346.

Local partnerships and private cooperative organizations known as mutual ditch companies represented the main organizational types that would replace the early speculative ventures. For Colorado this constitutes the first phase of sustained irrigation organization.[5]

The past one hundred years have witnessed the emergence of organizations designed to circumvent the obstacles to growth that were encountered by partnerships and private ditch companies. It would be incorrect to claim, however, that these smaller private organizations have been entirely eclipsed; indeed, in western Colorado unincorporated mutuals continue to be the dominant form of organization (figure 27). Nevertheless, the shift from smaller private organizations to larger public organizations, particularly in more water scarce regions, becomes important for understanding changes in water development patterns and practices.

Raphael Moses identified three phases in irrigation organization: an initial phase of private mutual ditch companies, a middle phase in which public irrigation districts become dominant, and a "modern era" of massive conservation and conservancy districts.[6] This progression was said to be driven by the need for greater financing and bonding powers to construct larger and more difficult projects. The power of a conservancy district to tax all land uses within its boundaries and to provide water services for non-agricultural uses represented an advance over the power of an irrigation district to levy assessments on irrigated land, which in turn exceeded in most cases the assessments made on shares held in mutual ditch companies. Explanations based on increased financing capability must be accompanied, however, by a recognition of Federal reclamation law, land reform efforts, and varying bargaining rules within different organization types. Before proceeding with competing interpretations of organizational change, it is useful to briefly review the full range of organization types and powers.

---

5. Utah represents a special case in which early community irrigation systems were later reorganized as larger private irrigation companies and water users' associations. In most other states early community irrigation systems and proportional sharing of risk during drought were the exceptions; small-scale private cooperatives and strict adherence to the priority doctrine were the rule. See George Thomas, *The Development of Institutions under Irrigation with Special Reference to Early Utah Conditions* (New York: Macmillan Co., 1920); James Hudson, *Irrigation Water Use in the Utah Valley, Utah,* Research Papers, no. 79, (Chicago: University of Chicago, Department of Geography, 1962); and Arthur Maass and Raymond L. Anderson, *. . . and the Desert Shall Rejoice* (Cambridge, MA: MIT Press, 1978), chapter 8.

6. Raphael J. Moses, "Irrigation Corporations," *Rocky Mountain Law Review* 32 (1960): 527-33.

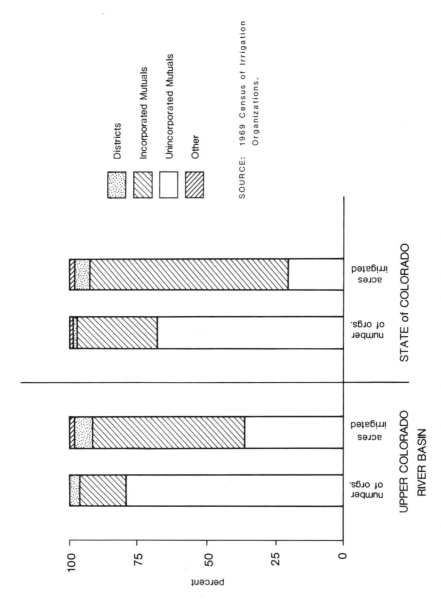

Fig. 27. Comparison of water organizations in Colorado and the Colorado portion of the Upper Colorado River Basin, 1969

Colorado statutes (C.R.S.) provide enabling legislation for at least eleven types of water organizations:
1. Ditch and Reservoir Companies, C.R.S. 7-42-101
2. Flume and Pipeline Companies, C.R.S. 7-43-101
3. Water Users' Associations, C.R.S. 7-44-101
4. Conservancy Districts-Flood Control, C.R.S. 37-2-101 thru 37-8-101
5. Drainage Districts, C.R.S. 37-20-101 thru 37-30-101
6. Irrigation Districts 1905, C.R.S. 37-41-101
7. Irrigation Districts 1921, C.R.S. 37-42-101
8. Irrigation Districts--1905 and 1921, C.R.S. 37-43-101
9. Internal Improvements Districts--1923, C.R.S. 37-44-101
10. Water Conservancy Districts, C.R.S. 37-45-101
11. Colorado River Water Conservation District, C.R.S. 37-46-101.

The fact that only three of these are found under Title 7 (Corporations and Associations) while the remainder come under Title 37 (Natural Resources--Special Districts) is not as illuminating as might be expected.

Simple distinctions between public and private water organizations misinterpret what is actually a complex array of organizational powers, rules, and duties. Mutual ditch companies, as quasi-private corporations organized under the Colorado Nonprofit Corporation Act, for example, are not subject to taxation; while irrigation districts, as public corporations with mainly private benefits, are subject to taxation; and water conservancy districts, as quasi-municipal corporations with a dominant public purpose, are not. Whereas the formation of a mutual is voluntary, formation of an irrigation district is based on the petition of a plurality, and formation of a conservancy district is based on a relatively smaller petition. Assessments are in each case set by a majority vote and are compulsory. This goes against the notion that individuals are free from compulsion in quasi-private corporations; it is true that they are not compelled to join. Water users' associations, quasi-private corporations organized to contract with the Federal government for reclamation project water, may levy assessments on other than a *pro rata* basis whereas mutual ditch companies may not. Representation in quasi-public and quasi-private corporations is by shares, whereas in quasi-municipal corporations it is democratic ("one man, one vote"). In irrigation districts assessments are not cumulative (members are not liable for the delinquency of other members), whereas in quasi-private and quasi-municipal corporations, they are. The intricacies of special district and corporation statutes would allow such comparison to continue at length. At issue are the attributes of water organizations that together constitute more of a continuum than a dichotomy between public and private entities. Novak proposed a classification of special districts based on power of assessment that provides a heuristic tool for proceeding (see table 21).[7]

The historical increase in financing capability identified by Raphael Moses provides only the skeleton of an explanation for the shift toward larger public forms of organization. Elwood Mead adds to this the ideological context of early irrigation district laws.[8]

> It is beginning to be realized that the waters of Western rivers are a great public resource which must be placed under public control in order to protect the public welfare. In order to do this, there must be reforms in irrigation laws and an assertion of a larger measure of public authority than has hitherto been thought necessary or desirable.

In recent years the emphasis of public authority has been on the resolution of conflicts among rival ditches and appropriators, but it has also sought to further settlement policies aimed at land reform.

The irrigation district as a quasi-public corporation was regarded as a remedy to the evils of speculative land practices and inefficiencies of overlapping small scale private ditch companies.[9]

> It was hoped in this way [through the Wright Act] to bring about the breaking up of large estates, or at least to prevent their owners from obstructing the development of other lands. [p. 210]

> The Colorado irrigation district act has been made use of in continuing under one management a number of rival and conflicting appropriations from the same stream, with a lessening of friction between water users and greater economy in the use of water, and lessened cost in the management of ditches [p. 382]

Transfer of speculative canal ventures to the resident irrigators was also regarded as an appropriate public goal, although Mead recognized the inefficiencies and fraud common in early district formation. In later years the formation of multipurpose water conservancy districts would also be justified as a response to the inequitable distribution of costs in irrigation development.[10]

> In most instances, the largest taxpayers in the area were contributing nothing to the cost of the works. In these rural areas, the economic life of the community varied directly with the prosperity of the farmers who were encumbering their lands to pay for the projects. As the water supplies improved, the area prospered and the merchants in the towns, the railroads and the utilities enlarged their volume and their profits. Out of this inequity grew the Colorado Conservation District Law.

---

7. Benjamin Novak, "Legal Classification of Special District Corporation Forms in Colorado," *Denver Law Journal* 45 (1968): 347-80; Charles J. Biese, "When Corporate Stock Becomes Real Estate," *Dicta* 21 (1944): 53-61; and Raphael J. Moses, "Irrigation Corporations," *Rocky Mountain Law Review* 32 (1960): 527-33.

8. Mead, *Irrigation Institutions*, pp. 347-348.

9. Ibid., pp. 210, 382.

10. Moses, "Irrigation Corporations," p. 531.

TABLE 21

SPECIAL DISTRICT CLASSIFICATION IN COLORADO

Regional Conservation Districts

1. Multi-county agency of the state created by statute
2. Public purpose
3. Power to levy ad valorem taxes and special assessments
4. Tax-exempt status
5. Absence of the power of local government

Example: Colorado River Water Conservation District

Quasi-Municipal Corporations

1. Independent corporate existence
2. Public purpose
3. Power to levy ad valorem taxes and special assessments
4. Tax-exempt status
5. Absence of the power of local government

Example: Water Conservancy Districts

Public and Quasi-Public Corporations

1. Independent corporate existence
2. Private benefits
3. Liability for taxes
4. Power to levy special assessment taxes
5. Territoriality

Examples: Irrigation districts except under the law of 1935, drainage districts, soil conservation districts

Quasi-Private Corporations (Non-Profit)

1. Independent or quasi-independent corporate existence
2. Private purpose
3. Tax-exempt except for some personal property (agricultural implements, livestock, and machinery)[*]
4. Power to levy assessments on shares

Examples: Mutual ditch companies, water users associations

Quasi-Private Corporations (For Profit)

1. Independent or quasi-independent corporate existence
2. Private purpose
3. Liability for taxes
4. Power to set water contract rates, subject to approval by county commissioners

Examples: water carrier companies, municipal water utilities

Private Corporations

"The day of the purely private water company operating for profit is gone from the field of irrigation as well as from the scene of domestic municipal supply."[**]

SOURCE: Adapted from Benjamin Novak, "Legal Classification of Special District Corporation Forms in Colorado," *Denver Law Journal* 45 (1968): 347-80.

[*] See *Logan Dist. v. Holt*, 110 Colo. 253, 133 P2d 530 (1943).

[**] Raphael J. Moses, "Irrigation Corporations," *Rocky Mountain Law Review* 32 (1960), p. 527.

Smith has shown in an interesting paper that redistributive forces operating within public water organizations result in inefficiencies in both total income and risk-bearing. He examines the notion that selection of one type of organization over another "depends upon an income comparison under private and public ownership . . . [and that], individuals vote on the basis of their economic interests," and he concludes that the strength of redistributive forces--as approximated by tenancy conditions, farm size, voting rules, and plurality requirements--aid in explaining the shift toward more public districts.[11] Institutional adjustments have proceeded in both directions, however, for as Moses notes, "Interestingly enough, many of the irrigation districts, once the works were paid for and the excess lands excluded, converted themselves to mutual irrigation companies."[12]

The idea that organizations were structured to take advantage of financing opportunites is clearly indicated in Colorado statutes enabling participation in Federal reclamation projects. In keeping with populist land settlement policies, the Reclamation Act of 1902 specified that,[13]

> No right to the use of water for land in private ownership shall be sold for a tract exceeding 160 acres to any one landowner, and no such sale shall be made to any landowner unless he be an actual bona fide resident on such land . . . (32 Stat. 388, sec. 5).
>
> Provided, that the right to the use of water acquired under the provisions of this act shall be appurtenant to the land irrigated, and beneficial use shall be the basis, the measure, and the limit of the right (32 Stat. 390).

In spite of these constraints aimed at curbing speculative claims, Federal reclamation law does not specify the types of organizations that may contract for water or participate in project development except again to demand minimum bonding capability. Consequently, water users' associations (quasi-private), irrigation districts (quasi-public corporations), and water conservancy districts (quasi-municipal corporations) whose articles of incorporation and by-laws meet the above requirements may all participate in Federal project development. Table 21 illustrates the range of organizational types participating in Federal projects in the study area. While it is correct to say that Federal financing has affected the relative size and scale of water

---

11. Rodney T. Smith, "Public and Private Agricultural Water Supply Districts," Center for the Study of Economy and the State, University of Chicago, 1980, p. 50. (Typewritten.)

12. Moses, "Irrigation Corporations," p. 530; and see also William R. Kelly, "Rehabilitation and Reorganization of Irrigation Projects that Parallel or Duplicate One Another: Legal Problems in Colorado," *Water Resources and Economic Development of the West* 7 (1958): 1-12.

13. 1902 Reclamation Act, 32 Stat. 388, sec. 5; 32 Stat. 390.

organizations, this by itself does not explain the shift toward more "public" organization types.

Organizational Change in Western Colorado

The roles of financing powers, Federal reclamation activity, land settlement policies, conflict resolution, and economic choice have been cited to explain aggregate shifts in organizational structure. Abrupt shifts in organization were not observed in the study area, however, perhaps reflecting the laggardly progress of water development and administration there. Mutual ditch companies have historically constituted the dominant vehicle for water development (figure 28). Incorporation of irrigation companies and districts follows the general pattern observed by Mead and Moses: rapid incorporation of quasi-private entities prior to 1900, some financed by foreign or eastern capital; a period of dissolution or reorganization into local unincorporated mutuals; formation of several large-scale entities organized to participate in Federal reclamation projects or more extensive local projects; followed in recent years by the formation of multi-purpose quasi-municipal conservancy districts (figures 29 and 30). These larger organizations, although fewer in number, embrace an increasingly important proportion of the irrigated acreage and capital invested in water development.

Close observation of individual cases reveals that the formation of an entity may be based on one or more of the explanatory forces cited. The Grand Valley Irrigation Company, the largest mutual in the study area, was formed through consolidation of several rival companies; the Grand Valley Water Users' Association was organized to participate in the Federal Grand Valley Project; the Orchard Mesa Irrigation District amended its by-laws but not its organizational structure in 1922 to enter into the Grand Valley Project; and the West Divide Water Conservancy District was structured to serve a range of water supply functions, including urban and industrial uses. Some efforts at reorganization, for example, the combined management of the Orchard Mesa and Garfield Gravity Divisions of the Grand Valley Project, have been judged unsuccessful and abandoned. Other organizational concepts, such as special subdistricts of the Colorado River Water Conservation District, mergers of water conservancy districts, canal company consolidation, and agro-urban water users associations, have yet to be tested.[14]

---

14. Evan C. Vlachos, et al., *Consolidation of Irrigation Systems Phase II: Engineering, Economics, Legal and Sociological Requirements,* Completion Report no.

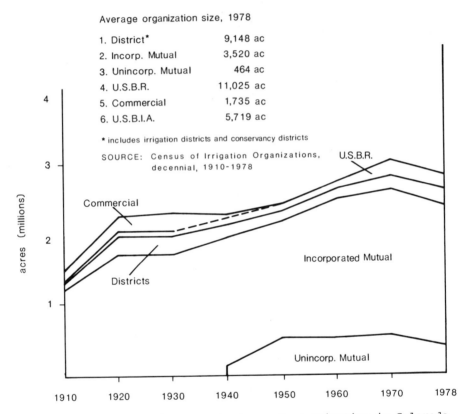

Fig. 28. Irrigated acreage by type of organization in Colorado 1910-1978

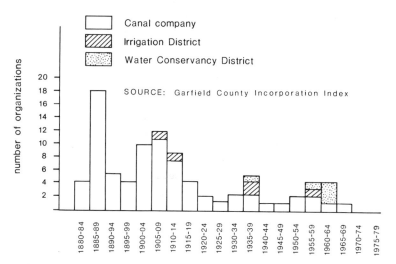

Fig. 29. Incorporation of water organizations in Garfield County

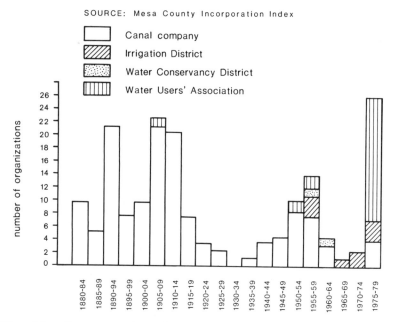

Fig. 30. Incorporation of water organizations in Mesa County

The issue for present purposes is not to identify the most important causative force in organizational change in the region; but rather to seek what is common in organizational adjustments of different sorts. Organizational change has served to resolve conflicts and accomplish the multiple purposes of interdependent entities. The resolution of conflicts among rival ditch companies through consolidation fifty years ago, for example, is of direct relevance to conflicts among urban and agricultural users in the Grand Valley today. In this sense, the process of organizational evolution described above clearly embodies the concept of integrated development outlined in chapter two. Changes in the jurisdiction and function of water organizations becomes important to the extent that it empowers the process of integrated development, that is, the establishment of functional linkages among water users.

Organizational Arrangements for Integrated Water Development

Chapters four through seven described the patterns and trends in water practice in western Colorado. Problems associated with current practice include chronic water shortages in District 45, excessive rates of application in most other areas, conflicts among urban and agricultural water users on Federal reclamation projects, uncertainity over the ownership and disposition of conserved water, speculative acquisition of agricultural water rights throughout the study area, and an absence of consensus on appropriate paths toward maximum water utilization. Four types of organizational adjustments which have potential application to local water supply problems throughout the study area are discussed below.

Type 1: Water Conservation on "Company Creeks"

The Roan and Parachute Creek drainages are almost wholly owned by energy firms which for the present time, are leasing land and water rights back to farmers at a nominal cost. In some cases the energy firms have stated an interest in continuing such leasing arrangements indefinitely, that is, except in places or at times when they will need to call in those rights for oil shale operations. Energy financed improvements in irrigation systems would not require special organizational adjustments, because as owners of both junior and senior rights the energy firms would benefit most from their own investment while maintaining or improving the viability of irrigation activities. The benefits of such investments accrue in the form of improved public relations and community development as well as more efficient water delivery.

---

94 (Fort Collins: Colorado State University, 1980).

As firms in both creeks intend to construct reservoirs, they might also consider multiple purpose reservoir management for agricultural uses in the creek valleys to establish and maintain their claims to beneficial use of appropriated water. "Company creek" water organization is predicated upon continued patterns of agricultural tenancy, the stability of which is yet to be determined. While improved water use could contribute to stabilizing rural land use, urban land markets and agricultural product markets will be determinative in specific locations. Given the constricted nature of oil shale development in Parachute Creek, the "company creek" concept would seem more plausible in the Roan Creek drainage. Control structures, measuring devices, ditch appurtenances, and in some cases, ditch linings, would ameliorate the highly variable rates of diversion and application along Roan Creek and its tributaries. To some extent, the District 70 water commissioner suggested that this process has already begun, as companies seek to protect their water rights by maintaining or improving ditch structures.

Type 2: Successive Reuse Agreements

Competition between urban and agricultural water uses will intensify as the population of the study area increases. In similar situations where urban uses would easily bid away water rights from irrigators, both uses have been maintained through arrangements for successive use and treatment of waste flows.[15] In western Colorado the flow path for such water might go as follows:
1. Collection of high quality tributary runoff for initial municipal and domestic uses
2. Land application of municipal wastewater
3. Industrial use of agricultural wastewater
4. Revegetation and dust control using industrial wastewater.

In addition to maintaining complementary water uses, total treatment costs may be reduced by planned reuse. Proposals for industrial use of naturally saline waters such as from Glenwood Springs include financial incentives to offset higher treatment costs.[16] Recalling the discussion in chapter three, however, a water

---

15. U.S. Environmental Protection Agency, *Northglenn Water Management Program, Final Environmental Impact Statement* (Denver, CO, 1980); and Musick and Cope, "Reclaiming Municipal Wastewater for Agricultural Use and Groundwater Considerations of Such Use," unpublished report (Boulder, CO, 1981).

16. Loretta C. Lohman, and J. Gordon Milliken, "Financial Incentives for Electric Utilities to Reuse Low Quality Waters in the Colorado River Basin," paper presented at the 1981 Water Reuse Symposium II (Washington, D.C., 1981); C. Earl Israelson, et al., *Use of Saline Water in Energy Development* (Logan, Utah: Utah Water Research Laboratory, 1980); and U.S. Department of the Interior, Bureau of Reclamation, *Saline Water Use and Disposal Opportunities,* draft report (Denver, CO, 1981).

right allows only a single use, except in the case of developed water. In at least one water rights transfer case in western Colorado, a stipulation to the decree attempts to preclude the reuse of municipal wastewater.[17] If, however, the amount transferred has already been diminished to the level of historic beneficial use, such stipulations would seem to have only a weak claim on the right to reuse.[18]

The organizational context of reuse could also bear on rules which constrain reuse. An ordinary appropriation of wastewater is subject to the call of the river. Intraorganizational exchanges of reservoir water and direct flow rights may be arranged without loss of priority if a suitable plan of augmentation or demonstration of "no-injury" can be shown.[19] Contracts for water delivery and treatment among different uses would probably best be facilitiated within a conservancy district framework, just as rental arrangements occur without interference among members of mutual companies and irrigation districts.[20] This type of arrangement deserves consideration in the modified West Divide proposal. The reduction in planned storage volumes; the projected growth of municipal, domestic, and industrial demands; and the increasing costs of financing wastewater treatment all argue for a plan of successive reuse if the conservancy district is to meet its intended multiple purposes.[21] If agricultural use of municipal wastewater were credited with the benefits of reduced treatment costs, the feasibility of the irrigation sector's participation in the project might be improved.[22]

---

17. Division 5, Water Court Case No. 79CW350 and No. 79CW351, water rights change and plan of augmentation for Battlement Mesa, Inc. new town. "This decree shall not be construed to allow reuse, recycling or successive use of water diverted under the priority for the 20 c.f.s. of the Dow Pumping Plant and Pipeline, which is the subject of Case No. 79CW350 for the purposes allowed herein." See also J. Gordon Milliken et al., *Overcoming Legal and Institutional Barriers to Planned Reuse of Water in the Colorado River Basin* (Washington, D.C.: U.S. Department of the Interior, Office of Water Research and Technology, 1979).

18. Under an unenforceable 1940 stipulation between Denver and several Front Range ditch companies, Denver was not to " . . . .use or attempt to use or lease any water, *irrespective of source,* which shall have been once used through its municipal water system . . . " in *City and County Law Review* 44 (1973), p. 312. Although the "developed water" argument could not be applied in western Colorado, other arguments could be directed toward overly restrictive stipulation on decrees.

19. Musick and Cope, "Colorado Water Law and Clean Water by Irrigation with Sewage Effluent," unpublished report (Boulder, CO, 1981).

20. Raymond L. Anderson, "The Irrigation Water Rental Market: A Case Study," *Agricultural Economics Research* 13 (1961): 54-58.

21. "West Divide (water conservancy district) Still Hopeful of Funding," *Rifle Tribune,* December 23, 1981; but see, "Bar 70 Challenges Yellow Jacket (water conservancy district) Water Policies," *Meeker Herald,* January 7, 1982.

22. U.S. Department of the Interior, Bureau of Reclamation, *Potential Modification in Eight Proposed Western Colorado Projects for Future Energy Development,* Special Report, chapter VI: West Divide Project (Washington, D.C., 1980).

The small volume of reservoir storage in the study area could result in junior storage rights held by energy firms becoming more reliable than their senior direct flow rights. Reuse agreements analogous to those in the Northglenn Project could be fashioned among industrial and urban uses, particularly in district 39. Transfer of municipal direct flow rights to reservoir storage on Elk, Roan, and Parachute Creeks could give greater security to municipal and domestic supplies. Industrial reuse of early priority municipal wastewater could improve the security of junior storage rights, again contingent upon plans of augmentation. The longevity of such arrangements would, of course, depend on the period for which reservoir design capacities will exceed industrial demand.[23]

Type 3: Multipurpose Water Development Organizations

Discussion of successive reuse arrangements drew into relief the uncertain status of traditional multipurpose water development within the study area. Many of the proposed projects have languished for several decades, and some conservancy districts face challenges to their continued existence. Lines of conflict have been clearly delineated among environmentalists, recreationists, and reclamationists.

Certainly the zenith of water development in the Upper Basin occured largely through the efforts of Congressman Wayne Aspinall of Grand Junction who served as the Chairman of the House Committee on Interior and Insular Affairs from 1958 to 1974. Much of this simple multiple purpose management was set, however, within the organizational structure of irrigation activity and Federal power development. The same forces that ushered Aspinall out of Congress in 1974, that reduced the political homogeneity of the Western Slope and its organizations such as Club 20, and that reduced agriculture to the margins of economic viability, have also broadened the meaning of multiple purpose water development in western Colorado.

This is nothing new in the state of Colorado where conflicting interests over the Colorado River headwaters are longstanding. In the earlier conflicts, however, resolution depended more upon the distribution of development benefits, such as storage reservoirs, than upon the type of development chosen.[24] The recent divergence of

---

23. Representatives of energy firms indicated that the volume of their present water rights holdings and planned storage facilities was based on some future median level of oil shale development and thus far exceeds the requirements of current projects. On the range of energy water requirements and synfuels industry scale forecasts, see Colorado Energy Research Institute, *Water and Energy in Colorado's Future* (Boulder, CO: Westview Press, 1980).

24. See, for example, Colorado Water Conservation Board, *The Green Mountain Reservoir Problem* (Denver, CO, 1960); and U.S. Department of the Interior, Bureau of Reclamation, *Final Environmental Statement: Colorado-Big Thompson Windy Gap*

interests in Western Slope water management has been accompanied by changing Federal policies toward project cost-sharing, leading most recently to the reorganization of administrative agencies for state water project financing.[25] These two trends call first, for reappraisal of traditional patterns of vertical integration that link special interests and Federal agencies; and second, for consideration of alternatives to achieve horizontal integration of water organizations and interests.

Just as consolidation of ditch companies enabled more effective local water management, several conservancy districts (Bluestone, West Divide, and Battlement Mesa) have considered a merger to pool their resources and link several of their common project goals. Such a merger would establish linkages among upstream and downstream areas as well as between the more plentiful supplies of Grand Mesa and the intensive demands of the West Divide district.

A second type of organizational adjustment would involve the formation of subdistricts of the Colorado River Water Conservation District. In theory, such a subdistrict could represent a broader range of water management interests than a repackaging of former water conservancy districts. A strong precedent exists for such subdistricts in the metropolitan subdistrict of the Northern Colorado Water Conservancy District which has planned and financed the Windy Gap extension of the Colorado-Big Thompson Project.[26] At present, the Colorado River Water Conservation District is not perceived as representing the full range of water management interests in western Colorado, given its overriding concern with the construction of additional storage facilities on the Western Slope; but this could change as the need increases for collaboration among local interests.

A final organizational vehicle for integrated development of a general character exists in voluntary associations of water interests. These presently include: Water for Colorado, the Colorado Water Congress, and Club 20/Colorado West. These organizations provide an alternative forum (to the courts) for discussion of water issues and have assembled a much wider range of interests than have the various water supply organizations.

Project, Colorado, Attachments, (Denver, CO, 1981).

25. Robert W. Miller, "Cost Sharing for Water Projects: Past, Present, and Future," paper presented to the Council of State Governments, Sacramento, CA, 12 June 1981.

26. C.R.S. 1973, 37-45-120.

Type 4: Organizational Solutions to the "Conserved Water" Problem

Legal restrictions on the use and transfer of "saved" water are often cited as constraints on conservation.27 Chapter seven raised this issue in connection with irrigation efficiency improvements made under the Grand Valley Salinity Project. In water rights transfer cases, the issue is traditionally resolved in terms of injury and historic beneficial use, though chapter six noted apparent inconsistencies in objections, stipulations, and administration under different local conditions. Recent policy proposals have concentrated on amendment of water transfer rules, establishment of public "water banking," and water brokerage. Water banking and brokerage have been implemented in some western states during severe water shortages, but have received only limited trial and review.28

In addition to these proposals, it should be recognized that the variety of organizations holding water rights implies a wider range of solutions to the conserved water problem than hitherto expected.29

> Water rights of an individual are quite specific in terms of the nature, place, timing, and amount of use...... On the other hand, water rights owned by municipalities or water conservancy districts allow a multitude of uses leaving the agent owner free to make and to manage the individual allocations for water use between and among constituents over time. Where water rights vary so greatly in terms of geographic extent, allowable uses, and the age, character, and operating latitude of the appropriator, it is clear that the resolution with which the State Engineer sees third party impacts to a particular water transfer differs vastly from one water right to another.

In general terms, then, there are internal and external types of solutions to the conserved water problem which apply to the study area as a whole and the Grand Valley area in particular.

Internal solutions are those in which the conserved water is put to some beneficial use by the conserver. In such cases conservation practices may enable more frequent irrigation, crop changes, longer irrigation sets, or improved scheduling and

---

27. George L. Radosevich and Gaylord V. Skogerboe, *Achieving Irrigation Return Flow Quality Control Through Improved Legal Systems* EPA 600/2-78-184 (Ada, OK: U.S. Environmental Protection Agency, 1978); Warren E. Bergholz, Jr., "Water Saved or Water Lost: The Consequences of Individual Conservation Measures in the Appropriation States," *Land and Water Law Review* 11 (1976): 435-457; and Willis H. Ellis and Charles T. DuMars, "The Two Tiered Market in Western Water," *Nebraska Law Review* 57 (1978): 333-367.

28. Sortirus Angelides and Eugene Bardach, "Water Banking: How to Stop Wasting Agricultural Water," (San Francisco: Institute for Contemporary Studies, 1978); and Jay M. Bagley, Kirk R. Kimball, and Lee Kapaloski, *Feasibility of Establishing a Water Rights Banking/Brokering Service in Utah* (Logan, UT: Utah Water Research Laboratory, 1980).

29. Ibid., p. 21. See also Vlachos, *Consolidation of Irrigation Systems Phase II;* and Raymond L. Anderson, "Organizational Arrangements in Water Transfers," *Water Resources and the Economic Development of the West* 19 (1961): 1-8. Anderson discusses the sale of irrigation company stock, rental of ditch and reservoir stock, transfer of Colorado-Big Thompson allotments, and Northern Colorado Water Conservancy District allotments. While common on the Front Range, these transfer arrangements are virtually absent on the Western Slope.

application. Internal solutions rarely invoke legal challenge, except for blatant "enlargements of use" such as canal extensions, irrigation of additional lands, storage of conserved water, and longer diversions. Litigation occurs most frequently when conservation improvements are followed by attempts to change an original water right.

It follows that the range of internal solutions available to an organization varies with its relative scale--measured in terms of water rights, service area, use, or membership. Larger organizations have more opportunities for internal reallocation both within and among uses. Statutory requirements and organization by-laws also affect the range of internal solutions. An irrigation company's shares of water may be deflated in value or transferred within the organization to distribute the benefits of conservation. A water conservancy district may reallocate water among its various decreed uses, but usually not outside its service area.[30] An organization drawing Federal project water, which is strictly appurtenant to the land, on the other hand, may only improve deliveries to insure that its members receive the prescribed duty of water as needed. Grand Valley Project organizations seek precisely this last benefit as a means of reducing conflicts among agricultural and suburban water users.

As a general conclusion, then, consolidation, merger, and expansion of irrigation organizations increase opportunities for internal solutions to the conserved water problem--but at what cost? This is a difficult question to answer. The *Shelton Farms* case identified the allocative system, that is the public, as losers if appropriation of salvaged water is permitted. Certainly a portion of the water rights queue may be injured by large-scale increases in beneficial use. Downstream users may be injured by altered return flows. Upstream users may have grounds for complaint against continued high diversion rates after implementation of water saving improvements, as in the case in the Grand Valley.

On the other side of the ledger, internal solutions have negligible public transaction costs and can actually facilitate efficient exchange among water users (within an organization) in ways that could serve as a model for intraorganizational transactions.[31] Past water practice thus endorses the selection of

---

[30]. The magnitude of such reallocation may be challenged, however, by other district beneficiaries, see "Bar 70 Challenges Yellow Jacket [water conservancy district] Water Policies," *Meeker Herald,* January 7, 1982.

[31]. Raymond L. Anderson, "The Irrigation Water Rental Market: A Case Study," *Agricultural Economics Research* 13 (1961): 54-58 presents a good case study of intraorganizational water transactions.

internal solutions, although the escalating trend in water rights disputes and the costs mentioned above may act to limit such arrangements in the future.32

External solutions to the conserved water problem involve either transfer, exchange, or abandonment of water rights as discussed in chapters four and seven. This class of solutions encounters many more obstacles and considerable transaction costs except in the case of voluntary water rights abandonment. Public water brokerage, perhaps through the Division Engineer's office, could reduce some of these costs and improve the data base for public water management; but water banking, in which the State has a "proprietary" interest in transfers, is out of keeping with Colorado's present practice of administrative restraint.33

In Grand Valley water managers argue that conservation improvements will lead to more efficient deliveries and not to waste or windfall. Assuming this is correct, it is still the case that more efficient deliveries are required to serve an increasingly less efficient, more numerous, and more heterogeneous assemblage of water users. Absolute water demand has not increased, but scheduling conflicts and operating requirements have, and for the delivery system to work as it has in the past--bankfull from April to October--all of the conserved water must continue to be diverted.

The problem is rooted in water management within the Grand Valley Project. The operating organizations resemble public water commissioners in that they administer water delivery only from the main canal to the farm laterals. Conflicts among the laterals are supposed to be resolved locally. The laterals are not organized to cope with the recent pace of land use changes so that conflicts often return to the major operating organizations. Voting privileges in some of the operating organizations are limited to landowners of one acre or more, in effect excluding the "new" users (though they are obliged to pay higher rates than the farmers per unit of water).34 Efforts to resolve recent interorganizational

---

32. Other obstacles to consolidation of water supply organizations are said to include: tradition, pride of possession or association with a specific ditch company, organizational rivalry, and conservatism among irrigators. See Ronald H. Strahle, "Problems in the Rehabilitation of Irrigation Projects Through Reorganization and Merger," *Water Resources and the Economic Development of the West* 6 (1957): 95-102. For these reasons, Strahle suggested that organizational change was only likely in cases of severe financial hardship or bankruptcy. Important but relatively unexplored economic obstacles lie in increased administrative costs to facilitate exchange, and in the distribution of economic benefits and costs to those involved in the reorganization.

33. Permit procedures, as in Utah, would probably expedite a centralized water banking system, see Jay M. Bagley, Kirk R. Kimball, and Lee Kapaloski, *Feasibility Study of Establishing a Water Rights Banking/Brokering Service in Utah* (Logan, UT: Utah Water Research Laboratory, 1980).

34. On Orchard Mesa, for example, assessments for irrigated land range from

conflicts have included formation of local suburban water users' associations and regulation of water system requirements for new subdivisions in the Grand Valley Project area.[35]

Consequently, any proposal for an external transfer of conserved water would immediately run up against the organizational and technological limitations of the existing delivery system, and it would have to consider the costs of adjustment in that system, such as changes in headgates or delivery practices. If this were not enough, strategic considerations impose further barriers. The principal beneficiaries of reduced diversions in Grand Valley would be transmountain diverters and not junior Western Slope water users. In fact, the juniors benefit from historically high diversion rights both at the Shoshone Power Plant and Cameo, and they would probably claim injury from any attempt to enforce a reduction in diversion rights. Similarly, transmountain diverters would actively oppose any attempt to exchange diversion rights on the Western Slope that might seem to circumvent the appropriation doctrine. Because Federal water is appurtenant to the land, an external solution involving Federal project water could only come in the form of water rights abandonment.[36] In the larger context of the study, external solutions could best be facilitated by a regional organization such as the Colorado River Water Conservation District which could serve as a broker among Western Slope water users or as a banker of its own supplies.[37]

Summary

Water resources are managed by a heterogeneous array of entities, the very structure of which can limit the effects of allocation rules. The internal structure and operating rules of an entity were shown to govern its flexibility for using existing water rights or conserved water. Those adjustments in organizational structure that internalize water exchange arrangements tend to facilitate solutions to problems of integrated development, e.g.

$19.86 to $32.80 per acre; residential and commercial lands are assessed, by contrast, at $178.34 per acre. From Orchard Mesa Irrigation District, *Report on Rehabilitiation and Betterment Program*: *Pumping and Conveyance Facilities* (Grand Junction, CO, n.d.), p. 53, and interview with Mr. Donald Caraway, Orchard Mesa Irrigation District, August 1981.

35. Mesa County, County Clerk's Office, "Corporation Index," Grand Junction, CO.

36. 32 Stat. 390. The Grand Valley Irrigation Company, which does not receive Federal water but does participate in the salinity control program would be an exception. More complicated issues arise with entities having supplemental water contracts with the Bureau of Reclamation based on a given duty of water. Whereas the Federal supplemental water would remain appurtenant to the land, other individual decrees might be transferable.

37. C.R.S. 1973, 37-46-107.

water conservation, successive reuse, and multiple purpose water management. Internal solutions to the conserved water problem, like the prior appropriation docrine, embody the principle that property rights holders are entitled to the fruits of their conservation efforts. Whether this will be accepted within Colorado water law will have to await more explicit legal definition of water diversion rights, beneficial use, and waste.

Organizational scale and structure were shown to affect the feasibility of conservation proposals. Organizational evolution toward larger and more complex entities in response to changing water management problems has thus exemplified the process of integrated water development in general terms. The idea that organization into larger public entities can contribute to conflict resolution, exchange, or adaptability must be added to a set of commonly held explanations that such developments enabled increased financing, fostered regional development objectives, and allowed redistributive forces to direct the path of water management. These ideas were reviewed for the study area in western Colorado and were found to be complementary rather than exclusive of one another. Organizational change has occurred in a disjointed fashion--rapid incorporation of private ventures in the late 19th century, irregular authorization of Federal projects, and only occasional formation of quasi-municipal entities--the resulting pattern becoming an overlay of organizational periods no one of which has yet risen to dominate the management of water in western Colorado.[38]

---

38. Although the issue of vulnerability to water supply hazards lies beyond our present scope of research, it should by no means be inferred that more complex water organizations are necessarily less vulnerable to water supply disruptions than decentralized or locally autonomous systems.

CHAPTER IX

CONCLUSION

The Colorado Water Study, a statewide assessment of water issues, nears completion after over five years of effort. Two recent remarks by officials responsible for the Water Study raise questions which provide a fitting preface to the conclusions of this monograph. First, the director of the Colorado Water Conservation Board states, "We have a free market system which recognizes private property rights in a property called water. You can trade them in the same way you might buy apples and oranges down at the store, and that's what this water study is all about."[1] The findings of the Water Study will presumably chart the general path of this alleged "free-market" system; describe specific regional trends; and highlight as problems various market inefficiencies and failures (e.g., resulting from inadequate definition of property rights to water, constraints on water transfers, inaccurate market signals, and so on).

A somewhat different perspective is provided, however, by the former director of the Colorado Department of Natural Resources who states that, " . . . unless you intervene in the process, water tends to flow toward money, and that creates a lot of social and economic problems you have to deal with."[2] In this case public policy problems seem to arise from the very operation of the water market. The opposition between these two statements should not be exaggerated, however, for they have common ground on the issue of water transfers. Water transfers curiously seem to represent the problem and the solution at the same time. How can this be?

---

1. J. William McDonald quoted in "Water Study Likely to Spark Use Debate," *Denver Post,* 19 July 1981.

2. D. Monte Pascoe, ibid.; and interview in Denver, Colorado, August 1981.

A common explanation for this paradox is that water transfers presently are made in an inefficient manner, and if they were made more efficiently, most water policy problems would be resolved. On the other hand, the issue might lie elsewhere--not in water transfers per se, but in the general relationship among Colorado water users. Colorado has free market mechanisms, to be sure, but it has few water markets to speak of and few public policies aimed at the development of regional water markets. Although the Colorado Supreme Court has boldly endorsed a path toward "maximum utilization," it too has offered little counsel as to the nature of that path. The path proposed in this reserach seeks more effective linkages among water users and water use areas.

Summary of Methodological Findings

The concept of integrated water development gives insight into what is meant by "the new drama of maximum utilization."[3] Integrated development was explicitly defined in this monograph as a process of adjustment in water use patterns and practices resulting from cooperative exchange among water users--as the active search for linkages among water users and for policies which foster such linkages. A geographical method of research was outlined to conduct the search for water use linkages. The method consists of the following tasks: describing the distribution of perceived problems; investigating the areal context (or domain) of those problems; "mapping" the range of choice; and finally, identifying new choices revealed through the process of search and mapping. The proposed method represents less a radical reformulation of research in water resources geography than a synthesis of various strands of geographical research, which White identified over twenty years ago as: resource estimates, technological assessments, studies of perception and range of choice, economic efficiency analyses, spatial linkages, and social guides.

This monograph began with a critique of recent Upper Basin water studies, observing the following: the failure to differentiate among regional patterns of surplus and scarcity; the neglect of adjustments in water rights distribution, water use patterns, and organizational arrangements; and finally, the premature dismissal of opportunities for improved water use efficiency. Geographical research was shown to offer a possible remedy for each of these shortcomings.

---

3. Colorado Supreme Court in *Fellhauer v. People* (167 Colo. 320, 447 P.2d 986, 1968).

The common pitfalls of regional water studies, however, have not yet been surmounted: these pitfalls being general confusion over what such studies can accomplish, and the conflation of different regional problems under a general river basin planning framework. In the first case, river basin studies were presumed to result in implementable plans and policy choices. This study, by contrast, represents regional water studies as, first, a hierarchical inventory of water control focusing on the distribution of water management problems; and second, as a search procedure for spatial linkages which may resolve those problems. Such investigations do not purport to identify an optimal solution or even to restrict their scope to the set of currently feasible solutions. Geographical investigations of this sort are for this reason logically antecedent to both engineering design and feasibility assessments.

The second error, conflation of regional problems, can only be avoided by altering the scale and scope of investigation to suit the problems under consideration. Most regional water studies, dating back to the early river basin surveys, have done so by describing the water system and its problems in hierarchical fashion. These inventories of water control in the Colorado River are generally referred to as "the law of the River." Rarely, however, are assessments of water management choices then merged with the detailed findings of preliminary surveys. In the search for "spatial linkages" among water management problems such merger is achieved. Thus, the proposed research method closely parallels the process of integrated water development.

Summary of Substantive Findings

Fifteen years of fervent concern over the adequacy of Upper Basin water supplies have recently given way to forecasts of regional water surpluses, at least until the turn of the century. This reversal of opinion demands an answer to the question: "If regional scarcity is not the problem, what are the real water management problems?"

Description of water management arenas and control points affecting water use in western Colorado helped identify the most important areas of perceived water scarcity. Foremost among these were the Lower Basin States and the Front Range of Colorado, both of which lie outside the study area. Western Coloradans are united in their apprehension about these transbasin and Lower Basin water seekers--to the point that inefficient water use and loose administration are tolerated in order to reduce apparent water surpluses in the region. Behavior which distorts regional water demand in this manner surely qualifies as a water management problem.

The contrast between the Upper Basin and Lower Basin water problems has many parallels in smaller units of the Colorado River system. The study area displayed a complex mosaic of surplus and scarcity which is typical of most areas in the Upper Basin. The Colorado River main stem, for example, ordinarily carries a considerable water surplus, while its tributaries are subject to annual shortages. Administration of tributary diversions involves regulating headgates to irrigation ditches, whereas administration of the main stem has historically involved regulation of transbasin diversions and reservoir storage releases. Main stem users, even though junior in water rights priority, are cushioned between large senior decrees upstream at Dotsero and downstream at Cameo. Although the separation between main stem and tributary water use in the area has not historically posed a problem, large-scale development of junior main stem diversions would necessitate clarification of currently ambiguous reservoir operating policies. Moreover, any increase in the frequency of water exchanges between the main stem and its tributaries would necessitate more coordinated administration.

Water districts were originally delineated for administrative convenience along watershed boundaries. As a result, water supply conditions tend to be quite heterogeneous within any given water district. Nevertheless, different problems appeared to be concentrated in each of the water districts. Water surpluses were inferred from the high irrigation application rates in district 39 and the Grand Valley area of district 72. In each of these areas reservoir storage and high relative seniority provide a reliable water supply throughout the growing season. Excessive diversion rates in district 70, on the other hand, indicated ineffective water control and measurement. Water scarcity, by contrast, prevails in distrct 45, where the most important water problem is said to lie in the lack of reservoir storage. It was shown that low irrigation application rates in district 45 are due to a combination of relatively large irrigable areas and small watershed yields. Again, what are the real water management problems here? The problem is not that absolute levels of surplus or scarcity vary over the region but that relative water supply differences of such magnitude can coexist in close proximity.

Tributary basin water supplies also vary considerably among themselves, but these differences pose few administrative or policy problems. Where intertributary diversions occur or where the stream network is complex, the prior appropriation rule establishes an effective administrative procedure. Aside from current administrative manpower deficiencies, these intertributary differences in water supply represent no special problem.

The prior appropriation doctrine also governs water allocation within any given basin, and it is well accepted within the region. Priority of right establishes the relative risk of scarcity among users. Does that mean that the junior appropriator is content to have his water supply cut off and to watch his crops wilt while his senior neighbor overirrigates? Definitely not, and the resulting conflicts raise a host of issues which are considered separately below in the discussion of water rights, diversions, public records, and organizational arrangements.

At the finest level of detail, relative differences in water supply were observed among water users in the same water service area. Risks have increased for users on the ends of ditches; in areas of recent suburbanization, particularly on Federal reclamation project lands where water remains appurtenant to specific tracts of land; and in service areas subject to intensive land speculation. Whereas individual appropriators accept the proper application of the prior appropriation doctrine, water users within any given organization such as an irrigation district expect equal sharing of risk.

Major problems have arisen for larger organizations which serve a rapidly changing mix of uses. Ditch riding duties have escalated, peak water demands have shifted, maintenance of laterals has declined, and the principle of timed water diversion is ignored. Irrigators state that new water users act as though "any water in the ditch is available for use." The new users feel that the traditional agricultural water use organizations willfully disregard the changes necessitated by urbanization as well as the principle that users who pay for water from an organization should have representation (e.g. the right to vote) in that organization. These represent pressing problems in urbanizing areas, but, with few exceptions, they are regarded by the State as private concerns.

The State faces a problem comparable to that of the larger water organizations, however, in that it only administers diversions of water from public water sources. It is not involved in the distribution or actual use of water within an organization after the water is diverted, provided only that those uses fall within the broad and ambiguous scope of the term "beneficial use." Because changes in water use are supposed to be recorded in the water rights tabulation, and because approval of water rights changes depend upon an accurate record of "historic beneficial use," numerous problems of public administration arise which are difficult to solve under the present system.

Public Records

Poor record-keeping underlies many of the other administrative problems found in the study area. Records of both water rights and water diversions contain so many omissions and inaccuracies as to have little value for either conflict resolution or regional water management. Fortunately, Water Division 5 has recently increased its efforts to correct and update the water rights tabulation. The State also sought a budget increase in 1982 for water commissioners and deputies in Division 5 to improve field administration of diversions. No progress has occurred, however, on water rights abandonment lists. Thus, even when corrected, the water rights tabulation will contain many inflated decrees and paper rights.

Water Rights

In what sense are paper rights a problem? After all, if the full amount of excessive senior decrees is not diverted, then the more junior users do not suffer. At the same time, the State of Colorado appears to have developed more of the water allocated to it under the Colorado River compacts. Problems arise when inflated decrees are sold at face value, but are dramatically reduced after intensive investigation reveals substantially lower "historic benefical use." Moreover, rational water management is impeded when there is confusion over real versus apparent water demand. Finally, many inflated decrees are exercised through excessive diversions and overirrigation in an attempt to establish beneficial use, and thus excessive decrees have contributed to problems of erosion, salinity, waste, and scarcity.

Water Diversions

Analysis of diversion records gave support to the argument that the prior appropriation doctrine reinforces inefficient patterns of water use. Record-keeping, a form of property rights protection, was most thorough for the larger more valuable water rights in the system. Senior water users naturally divert water for longer periods during the irrigation season and at higher rates per

unit of land. This does not constitute a problem except where
diversion rates exceed a reasonable duty of water (i.e., the
quantity of water used for a particular crop under conventional
practice). Average diversion rates of 2.25 acre-feet per acre in
district 45, for example, barely meet water requirements, and
justify the application of an allocative rule during periods of
shortage. Diversion rates exceeding five acre-feet per acre in all
other districts, by contrast, are relatively large in relation to
crop needs. Excessive diversion rates represent an important water
management problem, particularly when juxtaposed with areas of
scarcity nearby.

Where does the problem lie? Although the law protects vested
rights, it also prohibits "waste," demands "beneficial use," and
allows water transfers. Water diversion problems surface precisely
because these aspects of property rights are ambiguously defined.
Because diversion rights in excess of "historic beneficial use" are
non-transferable, little incentive exists to reduce diversions. How
can a path toward maximum utilization be laid out when vested rights
are not adequately defined, and there are disincentives to reduce
diversion rates?

In a peculiar twist to the general findings of this study, it
was ascertained that senior users did not always have the highest
rates of diversion. Junior water rights holders in district 70, in
contrast with those in districts 39 and 72, had the most excessive
recorded rates of diversion, apparently due to relatively
ineffective structural water control. In other cases, junior
appropriators subject to late season water shortages overirrigate
early in the season with the hope of increasing soil moisture
storage. Such water diversion practices—and the attending problems
of waste, erosion, nutrient leaching, and water pollution—are not
surprising given the absence of water markets which could alleviate
the risk of seasonal water shortages among junior appropriators.

Water Rights Changes

Although in theory even a minor change in water use should
require a change in water right, many small shifts in water use go
unnoticed, unchallenged, and unrecorded. Formal application for a
water rights change, on the other hand, often provokes high levels
of conflict. Approximately one-third of all tabulated water rights
changes in the past decade have been contested, and this figure does
not include proposed changes that were not successfully completed.
Comparison of formal water transfer rules with their application in

water court case records revealed that rules are not consistently applied. Rather, they seem to vary with the user groups, areas, and types of proposals under consideration.

Objections to water rights changes occurred most frequently in transactions involving municipal and industrial water users, senior irrigation rights, and water reuse arrangements. The legal concept of injury serves as the basis for most objections, and over time it has become sensitive to many more aspects of water transfers than when originally formulated. As the means for objecting to water rights changes have increased, so have transaction costs. In contrast with the increasingly sophisticated interpretations of injury, less progress has been made in giving basic dimensions to the concepts of "waste" and "beneficial use," perhaps due to the strength of vested rights. Thus, the combination of high transaction costs and fear of challenges to "beneficial use" has served to inhibit more creative water exchanges.

Changes in Water Diversion Patterns

In contrast with the active scene in water rights acquisition, water diversion patterns have displayed little change over the past fifty years. The most important changes have occurred first in Federal reclamation project areas where diversion and application rates have increased; and second, in urbanizing areas, where irrigated acreage has declined and application rates have increased. In fact, except in water-scarce areas of district 45, irrigation efficiency appears to be worsening rather than improving. Although few major trends were observed, annual water diversion rates were directly associated with annual variation in hay prices, farm revenue, and the previous year's livestock prices. On the whole, however, water diversion patterns and agricultural land uses have remained quite stable, illustrating that caution must be exercised in making inferences about changes in water demand from changes in water rights.

Conservation Practice

The notion of conservation varies widely among water users in western Colorado. Analysis of salinity control improvements in the Grand Valley area revealed that adoption of conservation measures was most frequent in areas of water shortage, conflict, or inflexibilty, e.g. on Federal reclamation project lands. The types of improvements made depended more on cost-sharing assistance levels and water delivery needs than on any inherent interest in saving water. Outside the Grand Valley area, Agricultural Conservation

Program funding was garnered principally by larger organizations on the Colorado River main stem. Smaller irrigators were either not able or not willing to participate to any significant extent.

Resistance to conservation improvements in the Grand Valley area gave way only under generous cost-sharing assistance and tentative assurances against the loss of rights to conserved water. But can such assurances be made compatible with the prior appropriation doctrine in Colorado? Colorado case law would suggest not, but no test on the scale of the Grand Valley salinity control project has yet occurred. Until some transferable property right can be assigned to conserved water, little incentive will exist to proceed toward maximum utilization. To those who argue that the transferable right to water should never exceed that portion of a right productively consumed in evapotranspiration, the following reply was offered: any use is fully consumptive with respect to all upstream juniors at all times; it is partially consumptive with respect to all downstream users who may enjoy the use of the water at some later time; it is non-consumptive only for senior water users, regardless of their location or time of use; but, given the importance of instream flows, there can be no such thing as a non-consumptive use in any pure sense.

Organizational Evolution

The most immediate prospects for integrated water development did not appear to lie in either major modification of the appropriation doctrine or in economic incentives for conservation. Even where the marginal value product of water ranges over an order of magnitude, as between irrigation and industrial uses, the transaction costs, uncertainities, and conflict associated with water transfers have impeded more efficient water use.

A more intriguing possibility for integrated development lies in the evolution of water organizations. Because the priority rule is administered only at diversions from public water sources, water organizations have considerable flexibility in allocating water among users once it has been diverted. The trend toward larger multiple purpose organizations represents in part a desire for greater flexibility in water use and exchange. The expansion of organizational scale and consolidation of water systems tend to promote equal sharing of risk and to significantly reduce transaction costs for water reallocation.

The following types of organizational adjustment were found to have specific relevance for local problems in the study area:

"company creek" associations composed of energy firms and irrigators on Parachute and Roan Creek; rejuvenation of water conservancy district plans in district 45 to include municipal and energy users as well as planned reuse of wasteflows; canal company consolidation; and formation of subdistricts of the Colorado River Water Conservation District to provide an organizational umbrella for water banking and brokerage. These proposals for organizational change could serve to internalize the water allocation process. Claims of injury would check major evasions of the priority rule, and minor transfers would be greatly facilitated. Effective operation of the prior appropriation rule would evolve concomitantly, however, if regional water organizations or the State were to serve as brokers for rights of unequal risk.

In summary, it should be noted that many of these proposals have surfaced previously at one time or another only to fall prey to judgements of infeasibility--political, legal, and economic. Most of the problems were recognized more than fifty years ago. The search for "spatial linkages" between problems and solutions, however, provides a catalyst for integrated water development. Unfortunately, as court battles over proposed water transfers become more protracted, the prospects for improving water management seem more costly and remote. Organizational evolution was found to offer some immediate relief, but property rights refinements must also advance to provide a solid foundation for more efficient water use. These advances do not proceed, however, *solely* from the coordination of localized problems and choices. They also depend upon a fundamental shift--away from conflict and toward cooperation--away from complacency with free-market mechanisms and toward the establishment of true forums for water policy formulation and water exchange.

APPENDIX

TABLE 22

SUMMARY OF DECREED WATER RIGHTS
(Including the Colorado River)

Decree Type	All Streams n=52	District No.			
		39 n=8	45 n=13	70 n=8	72 n=23
No. AFLOW	2,960	708	764	294	1,194
Vol. AFLOW (c.f.s.)	14,531	2,102	1,557	742	10,130
No. CFLOW	526	113	173	34	206
Vol. CFLOW (c.f.s.)	11,089	5,306	1,388	853	3,542
Vol. ASTOR (a.f.)	86,875	22,816	2,096	524	61,439
Vol. CSTOR (a.f.)	969,230*	412,786	24,262	397,644	134,538
Mean Size AFLOW (c.f.s.)	4.91	2.97	2.04	2.52	8.48
Mean Size CFLOW (c.f.s.)	21.08	46.96	8.02	25.09	17.19

Source: Colorado Division of Water Resources, Water Rights Tabulation (1978). (Stream-Alpha printout.)

*Includes double counting of separate adjudications for Una Reservoir in districts 39 and 70 (decreed volume = 195,983 acre-feet).

No. AFLOW	=	Number of absolute flow rights
Vol. AFLOW	=	Sum of absolute flow rights (cubic feet per second)
CFLOW	=	Conditional flow rights
ASTOR	=	Absolute storage rights
CSTOR	=	Conditional storage rights
AFLOW/Mi^2	=	Sum of absolute storage rights ÷ basin area (similarly CFLOW/Mi^2, ASTOR/Mi^2, and CSTOR/Mi^2)
Length	=	Stream length
Area	=	Basin area (square miles)
Precip.	=	Annual average precipitation (inches)
Slope	=	Stream slope (feet per mile)
Yield	=	Annual basin water yield (a.f.)
Distance	=	Distance from stream mouth to nearest municipality
Year	=	Year of initial appropriation
Period	=	Initial period of sustained appropriation
Population	=	Population of nearest municipality
Demand	=	Population ÷ Distance

TABLE 23

MEAN VALUES OF BASIN WATER RIGHTS VARIABLES BY DISTRICT

Untransformed Variables	All Streams n=52	Without Colo. R.* n=48	District No.			
			39 n=7	45 n=12	70 n=7	72 n=22
Length (mi.)	17	15	14	12	20	14
Area (sq. mi.)	91	75	90	34	127	77
Precip. (in.)	22	23	25	21	21	25
Slope (ft./mi.)	346	374	440	416	183	391
Yield (a.f.)	96,643	91,506	123,364	42,114	135,286	94,382
Distance	5.9	6.4	1.5	4.9	10.0	7.5
Year	1,889	1,889	1,883	1,890	1,884	1,891
Period (yr.)	30	29	33	27	18	32
Population	5,400	4,311	2,691	1,590	419	7,594
Demand	16,984	3,018	7,746	380	685	3,694
AFLOW (c.f.s.)	277	124	219	108	100	109
CFLOW (c.f.s.)	213	77	208	99	37	37
ASTOR (a.f.)	1,662	1,793	3,250	139	75	2,778
CSTOR (a.f.)	18,601	8,315	25,028	1,855	18,596	3,250
AFLOW/Mi^2	2.9	2.8	2.4	3.7	1.4	2.8
CFLOW/Mi^2	2.1	1.2	2.5	2.3	0.4	0.6
ASTOR/Mi^2	22.4	24.3	23.0	5.8	1.6	42.1
CSTOR/Mi^2	262.5	93.0	228.5	18.4	40.6	107.1

SOURCE: Colorado Division of Water Resources, Water Rights Tabulation (1978)

*Colorado R. refers here to the Colorado River main stem.

TABLE 24

CORRELATION COEFFICIENCES FOR ABSOLUTE FLOW RIGHTS BY DISTRICT (AFLOW)

Variable	All Streams n=52	Without Colo. R. n=48	District No.			
			39 n=7	45 n=12	70 n=7	72 n=22
Length	.5447**	.4717**	.6449	.8774**	.3759	.4199+
Area	.7111**	.5010**	.5571	.9672**	.4355	.5903**
Precip.	-.2989*	.3740**	.3992	.4382	-.0911	.4375*
Slope	-.4989**	-.3560*	-.5493	-.3700	-.3792	-.3714+
Yield	.3293*	.5800**	.6018	.8188**	.3790	.5703**
Distance	-.1871	-.1666	-.3768	-.3392	.2127	-.0959
Year	-.2725+	-.4230**	-.5880	-.5806*	-.0242	-.3220
Period	.4976**	.3599+	.0312	.6703*	.2908	.4704*
Population	.5790**	.2120	-.1437	-.1286	−	.3979+
Demand	.4486**	.2956*	.6028	.1320	.2827	.1578

+α = .10; *α = .05; **α = .01; d.f. = n-2; two-tailed t test

TABLE 25

CORRELATION COEFFICIENTS FOR AREA-ADJUSTED ABSOLUTE FLOW RIGHTS BY DISTRICT (AFLOW/Mi^2)

Variable	All Streams n=52	Without Colo. R. n=48	District No.			
			39 n=7	45 n=12	70 n=7	72 n=22
Length	-.1782	-.5014**	-.0853	.1174	-.3677	-.6794**
Area	-.1318	-.3948**	-.0229	.0432	-.3813	-.4691*
Precip.	.3157*	.4873**	.7047+	.6098*	.2661	.5643**
Slope	.2001	.5782**	.2782	.4467	.2077	.7098**
Yield	-.2796*	-.3666*	-.0636	.1264	-.4354	-.4968*
Distance	.0016	.0525	.1320	.2706	.8494*	.0075
Year	-.2687+	-.2237	-.8692*	-.4672	.1037	-.1011
Period	.2407+	.1609	-.1441	.8104**	.0299	.0715
Population	.0460	-.1005	.1710	-.3160	–	-.1482
Demand	-.0311	-.1700	.1837	-.4098	-.5279	-.2874

+α = .10; *α = .05; **α = .01; d.f. = n-2; two-tailed t test

TABLE 26

CORRELATION COEFFICIENCES FOR CONDITIONAL FLOW RIGHTS BY DISTRICT (CFLOW)

Variable	All Streams n=52	Without Colo. R. n=48	District No.			
			39 n=7	45 n=12	70 n=7	72 n=22
Length	.6070**	.3676*	.3855	.8146**	.6700+	.5736**
Area	.4688**	.3170*	.2862	.9575**	.8044*	.5657**
Precip.	-.3471*	.2508+	.6842+	.3665	.2658	.1787
Slope	-.6370**	-.2159	-.2000	-.3616	-.3596	-.5185*
Yield	.3210*	.4473**	.4225	.7455**	.7921*	.6557**
Distance	-.2981*	-.2695+	-.3497	-.3712	.6651	-.3860+
Year	-.2362+	-.3589*	-.0137	-.4943	-.3338	-.2471
Period	.3138*	.2723+	.1463	.5841*	.8342*	.6397**
Population	.3352*	-.0334	-.6869+	-.1418	–	.1946
Demand	.5049**	.1394	-.4415	-.1401	.8154*	.5237*

+α = .10; *α = .05; **α = .01; d.f. = n-2; two-tailed t test

TABLE 27

CORRELATION COEFFICIENTS FOR AREA-ADJUSTED CONDITIONAL FLOW RIGHTS BY DISTRICT (CFLOW/Mi^2)

			District No.			
Variable	All Streams n=52	Without Colo. R. n=48	39 n=7	45 n=12	70 n=7	72 n=22
Length	.0910	-.1516	-.0708	.0737	.2228	-.2000
Area	.0527	-.0586	-.0520	.2552	.2868	-.1276
Precip.	.0706	.3209*	.5925	.6552*	.6881+	.3510
Slope	-.0828	.3687**	.2522	.4498	.0257	.2716
Yield	-.0695	-.1312	-.0086	.2617	.3288	-.0856
Distance	-.3753**	-.3064	.5657	-.0323	-.2719	.1772
Year	-.2594+	-.1572	-.0137	-.6175*	-.1429	-.2038
Period	.3348*	.3497*	-.0517	.6335*	.8724*	.3186
Population	.0831	.0090	-.4810	-.0772	-	-.0476
Demand	.2792*	.1457	-.6612	-.0423	.2852	.0497

+α = .10; *α = .05; **α = .01; d.f. = n-2; two-tailed t test

TABLE 28

CORRELATION COEFFICIENTS FOR ABSOLUTE STORAGE RIGHTS BY DISTRICT (ASTOR)

			District No.			
Variable	All Streams n=52	Without Colo. R. n=48	39 n=7	45 n=12	70 n=7	72 n=22
Length	.2156	.4363**	.4280	.4269	-.0991	.5829**
Area	.3937**	.5652**	.3030	.5292+	-.2884	.8290**
Precip.	.1910	.1763	.2802	.4867	-.3709	.0691
Slope	-.0493	-.2210	-.3704	.0025	-.1133	-.3543+
Yield	.3762**	.4104**	.3616	.5531+	-.4292	.5615**
Distance	-.0670	-.0993	-.3436	-.4623	.2114	-.0798
Year	-.2761*	-.2906*	-.6280	-.1436	-.3799	-.4360*
Period	.4060**	.4792**	-.1973	.3978	-.5491	.5782**
Population	.4630**	.5890**	.1538	-.0432	-	.6728**
Demand	.2448+	.3656*	.6516	.1431	-.2770	.2700

+α = .10; *α = .05; **α = .10; d.f. = n-2; two-tailed t test

TABLE 29

CORRELATION COEFFICIENTS FOR AREA-ADJUSTED ABSOLUTE STORAGE RIGHTS BY DISTRICT (ASTOR/Mi^2)

Variable	All Streams n=52	Without Colo. R. n=48	District No. 39 n=7	District No. 45 n=12	District No. 70 n=7	District No. 72 n=22
Length	.1045	.1704	.6653	.2342	-.1867	.2416
Area	+.0766	.0663	.5577	.2572	-.3602	.1758
Precip.	.1794	.1608	.6017	.4555	-.4625	-.0248
Slope	.0632	-.0183	-.5082	.1524	-.0313	-.1304
Yield	.2160	.2121	.5747	.2627	-.4643	.4191+
Distance	-.0823	.1185	-.4961	.0491	.1939	-.4886*
Year	-.0796	-.0390	-.6647	.0352	.1267	-.7003**
Period	.3141*	.3260*	.0748	.2386	-.7422+	.5183*
Population	.2334+	.2550+	-.1370	-.0029	-	.2504
Demand	-.1561	.2216	.6732+	-.0728	-.3225	.1081

+α = .10; *α = .05; **α = .01: d.f. = n-2; two-tailed t test

TABLE 30

SUMMARY STATISTICS FOR DISTRICT WATER DIVERSIONS, 1980

VARIABLE	Statistic	All Streams	Dist. 39	Dist. 45	Dist. 70	Dist. 72
Length of diversion (days)	mean s.d. c.v.	113.5 (73.1) .644 n=390	162.1 (67.7) .417 n=89	76.3 (67.0) .878 n=117	117.6 (65.4) .556 n=55	111.8 (65.6) .587 n=129
Mean Irrigation rate (a.f./ac.)	mean s.d. c.v.	6.67 (9.05) (1.36) n=342	11.65 (7.84) (.673) n=71	2.91 (3.23) (1.11) n=113	13.54 (17.44) (1.29) n=51	4.07 (2.62) (.644) n=107
Area-Adjusted Diversion Rate (c.f.s./mi.^{2})	mean s.d. c.v.	1.048 (1.099) (1.049) n=43	.623 (.355) (.570) n=7	1.559 (1.461) (.937) n=10	.364 (.209) (.574) n=7	1.11 (1.147) (1.033) n=15
Total Diversion Rate (c.f.s.) Incl. Colorado R.		48.3	439.2	421.4	292.8	2294.9
Total Acreage Incl. Colorado R.		163,850	16,497	31,414	9,674	106,265
Overall Irrigation Rate (AF/AC)		5.26	7.40	2.25	8.14	5.55
% of Structures Recorded		49.1%	47.9%	48.9%	67.7%	41.0%

s.d. = Standard deviation; c.v. = coefficient of variation

SOURCE: Water commissioners' annual reports for 1980.

NOTE: Mean irrigation rate refers to the average of individual ditch records for the volume of water diverted per acre. Overall irrigation application rates, on the other hand, were obtained by dividing the total volume of water diverted by the total number of acres irrigated in each district.

TABLE 31

CORRELATION COEFFICIENTS FOR PERCENTAGE OF
STRUCTURES RECORDED (%REC)

Variable	All Streams n=43	Without Colo. R. n=39	District No.			
			39 n=7	45 n=10	70 n=7	72 n=15
Length	-.2017	-.1043	-.6606	.2184	-.1839	-.4876+
Area	-.2092	-.1298	-.4569	-.0648	-.2784	-.4130
Precip	.0406	-.0695	-.0474	-.0107	.2618	-.0354
Yield	-.1960	-.1504	-.7414+	.0326	-.1361	-.5514*
Distance	.4128**	.3939*	-.0291	.6541*	.6265	.2933
Year	-.0798	-.1317	-.4325	-.3758	-.1051	.4074
Period	-.2318	-.1771	-.2504	.0910	.2386	-.3946
Demand	-.3171*	-.2732+	.0581	-.2507	-.4596	-.5316*
DAYS	.2255	.3548*	.8774**	-.2220	.4011	-.1402
AF/AC	.2397	.2854+	-.0819	-.0534	.3676	-.3732
MRANK	-.1346	-.1461	-.4683	-.2635	.7630*	.3618
CFS	-.1766	-.1163	-.3961	.1090	.0291	.1051
AFLOW	-.1676	-.0591	-.2911	-.0260	.5149	-.1089

+α = .10; *α = .05; **α = .01; d.f. = n-2; two tailed t test

```
Length   = Stream Length
Area     = Basin area
Precip.  = Average annual precipitation
Yield    = Basin water yield (a.f.)
Distance = Distance from stream mouth to nearest municipality
Year     = Date of initail decree
Period   = Initial Period of appropriation
Demand   = Population ÷ Distance
DAYS     = Mean length of diversion (days)
AF/AC    = Irrigation application Rate
MRANK    = Mean basin rank
CFS      = Sum of water diversions (c.f.s.)
AFLOW    = Sum of absolute flow rights
```

TABLE 32

CORRELATION COEFFICIENTS FOR TOTAL DIVERSIONS (CFS)

Variable	All Streams n=43	Without Colo. R. n=39	District No.			
			39 n=7	45 n=10	70 n=7	72 n=15
Length	.5553**	.4055*	.8286*	.7985*	.8632*	.2061
Area	.7048**	.3643*	.8108*	.8495**	.9051**	.1958
Precip	-.3227*	.5781**	.6349	.5906+	-.1740	.6746**
Slope	-.5218**	-.3251*	-.6543	-.1357	.7226+	-.4840+
Yield	.3435*	.5709**	.8132*	.7746**	.8422*	.4550+
Distance	-.1732	-.0582	-.3592	-.0268	-.4042	.2128
Year	-.2555+	-.3224*	-.4369	-.7721**	-.3957	.0217
Period	.4764**	.1956	.3994	.7558*	.5608	.0667
Population	.5458**	-.0268	-.5896	-.2713	-	-.0274
Demand	.4307**	.0951	.3800	-.1052	.8085*	-.2522
MRANK	.0822	.1033	.2253	-.1463	-.1692	.2292

+α = .10; *α = .05; **α = .01; d.f. = n-2; two-tailed t test

TABLE 33

CORRELATION COEFFICIENTS FOR AREA-ADJUSTED DIVERSIONS
(CFS/Mi^2)

Variable	All Streams n=43	Without Colo. R. n=39	District No.			
			39 n=7	45 n=10	70 n=7	72 n=15
Length	-.1957	-.4622**	-.1438	-.3370	-.6405	-.5383*
Area	-.2094	-.4370**	.0651	-.3677	-.6506	-.5202+
Precip.	.1538	.3598*	.7703*	.4955	.5151	.3688
Slope	.0805	.4824**	.3715	.7282*	.5236	.3460
Yield	-.3614*	-.4128**	-.0707	-.3680	-.7438+	-.3688
Distance	.0544	.2199	.1882	.6156+	.7724*	.3635
Year	-.0452	-.0647	-.7343+	-.2490	.0847	.1921
Period	-.0452	.0076	.1834	.5072	.0162	-.2119
Population	-.2177	-.3760*	-.0629	-.4725	-	-.6660**
Demand	-.2345	-.4042*	-.0370	-.6931*	-.6964+	-.7556**
MRANK	-.0879	-.1516	-.5462	-.6315+	.4753	.1759

+α = .10; *α = .05; **α = .01; d.f. = n-2; two-tailed t test

TABLE 34

CORRELATION COEFFICIENTS FOR IRRIGATED ACREAGE (AC)

Variable	All Streams n=43	Without Colo. R. n=39	District No. 39 n=7	District No. 45 n=10	District No. 70 n=7	District No. 72 n=15
Length	.5675**	.4905**	.8139*	.8383**	.9364**	.4596+
Area	.7148**	.3872*	.8047*	.9365**	.9154**	.4514+
Precip.	-.3489*	.2261	.6474	.4706	-.1025	.1360
Slope	-.5026**	-.2251	-.6312	-.2568	-.8109*	-.3778
Yield	.3489*	.4888**	.8019*	.7935**	.8942**	.5510*
Distance	-.2983+	-.4778**	-.3362	-.6266+	-.4906	-.5023+
Year	.4210**	.0595	.2749	.0553	.8134*	-.1278
Period	.5130**	.3686*	.4587	.6695*	.5848	.5797*
Population	.5483**	.0102	-.6701+	-.1678	-	.1018
Demand	.9543**	-.0042	.3729	-.1154	.9001**	-.2417
MRANK	.0547	-.0176	.2739	.0075	-.2577	-.2495
AFLOW	.9854**	.5903**	.6094	.9858**	.5080	.4355
CFS	.9854**	.5848**	.9751**	.9637**	.9595**	.3557

$+\alpha = .10$; $*\alpha = .05$; $**\alpha = .01$; d.f. = n-2; two-tailed t test

TABLE 35

CORRELATION COEFFICIENTS FOR LENGTH OF IRRIGATION SEASON
(DAYS)

Variable	All Streams n=43	Without Colo. R. n=39	District No. 39 n=7	District No. 45 n=10	District No. 70 n=7	District No. 72 n=15
Length	.2118	-.0208	-.5722	.0290	.4508	.0426
Area	.3716*	.1008	-.2357	-.0426	.4126	.0093
Precip.	-.1342	.1667	.1567	.4456	-.6085	-.3092
Slope	-.1970	.2710+	.6773+	.4772	-.5728	.1193
Yield	.1606	.0724	-.5582	.1924	.4234	-.0863
Distance	-.3303*	-.2832+	.0172	.0944	.0594	-.6209*
Year	-.3467*	-.3681*	-.5090	-.1391	-.0741	-.6041*
Period	.2609+	.1005	.1043	.1217	-.1739	.0944
Population	.3862*	.1866	.2671	.6106+	.2946	.3726
Demand	.5517	.5035**	-.0025	.3197	.2788	.6741**
MRANK	-.1651	-.3199*	-.8488*	-.5416	.0706	-.4784+
AFLOW	.4749**	.0378	-.1685	-.0734	.8004*	-.1969
CFS	.4695**	-.0498	-.1859	-.1373	.5350	-.3264
ASTOR	.0455	.0910	-.1310	.5136	-.2079	.0159

+α = .10; *α = .05; **α = .01; d.f. = n-2; two-tailed t test

TABLE 36

CORRELATION COEFFICIENTS FOR IRRIGATION APPLICATION RATES
(AF/AC)

Variable	All Streams n=43	Without Colo. R. n=39	District No.			
			39 n=7	45 n=10	70 n=7	72 n=15
Length	.2241	.1255	.1259	-.3317	-.3507	-.0038
Area	.1887	.1630	.3912	-.4737	-.2594	.0338
Precip.	-.2800+	-.0942	.6045	-.1013	-.0451	-.5490*
Slope	-.2801+	-.1223	.0326	.4343	.2600	.2449
Yield	.0985	.1010	.4407	-.4158	-.2896	-.0506
Distance	.1111	.2112	.1562	.2915	.5058	-
Year	-.1322	-.1700	-.3661	.4963	.7767*	-.3228
Period	-.0188	-.0582	.5826	-.0739	-.3024	.1858
Population	.0535	-.0060	-.3937	-.2396	-	.1992
Demand	.0969	-.0505	-.1261	-.4240	-.3101	.3182
MRANK	-.1527	-.2837+	-.3989	-.3498	.5453	-.4794+
AFLOW	.0922	-.0306	.3105	-.4733	.4655	-.4424+
CFS	.0853	-.1176	.4939	-.5056	-.2120	-.6090*
ASTOR	-.0065	.0210	.1381	.0538	-.3999	-.0203
DAYS	.4888**	.5038**	.3925	.1954	.4276	.5592*

+α = .10; *α = .05; **α = .01; d.f. = n-2; two-tailed t test

TABLE 37

CORRELATION COEFFICIENTS FOR SENIORITY (RANK), 1980

Variable	All Streams $n_1=389$ $n_2=342$	District No.			
		39 $n_1=89$ $n_2=71$	45 $n_1=116$ $n_2=113$	70 $n_1=55$ $n_2=51$	72 $n_1=129$ $n_2=107$
Diversion Rate (CFS)[a]	.0396	-.2179*	-.1676	.2219	.0013
Diversion Volume (AF)[a]	.0296	.2879**	-.1954*	.2229	.0144
Irrigated Acreage (AC)[b]	.0154	-.1655	-.0646	-.0625	-.0086
Application Rate (AF/AC)[b]	-.0520	.0234	-.1435	.5407**	-.3146**
Length of Season (DAYS)	-.2130**	-.1443	-.4375**	.0442	-.4697**

SOURCE: Water commissioners' annual reports for 1980.

[a] Sample size = n_1; all water diversion records
[b] Sample size = n_2; irrigation records only

TABLE 38

CORRELATION COEFFICIENTS FOR ALL WATER RIGHTS CHANGES
(ALLCHG)

Variable	All Streams n=43	Without Colo. R. n=39	District No.			
			39 n=7	45 n=10	70 n=7	72 n=15
Area	.4258**	.3419*	.8001*	.0433	.9450**	.3168
Precip.	-.3483*	.1113	.2376	.3899	-.2868	.1750
Yield	.4156**	.3965*	.8179*	.0912	.8504*	.2461
Distance	-.3858*	-.2742+	.0106	.3777	-.7848*	.0403
Year	-.3177*	-.4056*	.2031	-.4131	.2611	.0327
Period	.1936	.0456	.6063	.3485	.3621	-.0054
Population	.1645	-.1030	-.9792**	.3865	.9764**	.1387
Demand	.5387**	.1957	-.2270	.1430	.9522**	-.0192
Gross Transfers	.3404*	.8433**	-.9106**	.7800**	.9554**	.3672
% REC	-.2230	-.1705	-.4446	.3516	-.5020	.0971
AF/AC	.2937+	.2121	.5328	-.0753	-.3623	-.0621
MRANK	.1143	-.2445	.3554	-.3750	-.5657	-.0327
CFS	.3567*	.3867*	.6721+	.2779	.8290*	.4500+
AFLOW	.3434*	.2687	.2785	.1864	.2993	.5129+
CFLOW	.5443**	.4522**	.6326	.0848	.8005*	-.0409
ASTOR	-.0674	.0078	-.205	-.1496	-.1406	.3426
CSTOR	.4806**	.6642**	.9034**	.0580	.9604**	.1278

+α = .10; *α = .15; **α = .10; d.f. = n-2; two-tailed t test

TABLE 39

CORRELATION COEFFICIENTS FOR CONTESTED WATER RIGHTS CHANGES
(CONCHG)

Variable	All Streams n=43	Without Colo. R. n=39	District No.			
			39 n=7	45 n=10	70 n=7	72 n=15
AF/AC	.2654+	.2256	.4373	-.0444	-.1056	.4195
MRANK	-.0784	-.3665*	-.0981	-.1629	-.4903	-.4330
CFS	-.0215	.3474*	.6121	.5222	.7122+	-.1564
AFLOW	-.0215	.3667*	.4637	.6181+	.3037	-.1590
CFLOW	.1853	.2915+	.0036	.5504+	.6310	-.1463
ASTOR	-.0163	.0354	.2550	.2392	-.3254	-.1321
CSTOR	.3000+	.5433**	.5090	.6223+	.8800**	-.1269
Yield	.3944**	.4509**	.7303+	.5875+	.8755**	-.2284
Year	-.2819+	-.3892*	.0001	-.4352	-.0001	-.2479
Period	.0242	.0359	.8223*	-.3314	.1739	-.0839
Demand	.4213**	.3242*	.4229	.3238	.9518**	.1689
Gross Trans	-.0542	.6092**	.5235	.3563	.8776**	-.1269
% REC	-.1550	-.1382	-.2973	.2482	-.4298	.0467

+α = .10; *α = .05; **α = .01; d.f. = n-2; two-tailed t test

TABLE 40

MEAN VALUES FOR ANNUAL DIVERSION RECORDS AND INDEPENDENT VARIABLES
(n = 18)

VARIABLE	MEAN	STD. DEV.	C.V.
Water Diversion			
39 AF	120,808	22,303	.185
39 AC	20,225	3,465	.171
39 AF/AC	6.14	1.46	.238
45 AF	72,820	18,880	.259
45 AC	24,226	5,689	.235
45 AF/AC	3.10	.80	.258
70 AF	42,833	25,236	.589
70 AC	8,613	1,971	.229
70 AF/AC	4.85	2.26	.466
72 AF	909,788	239,111	.263
72 AC	132,949	16,427	.124
72 AF/AC	7.00	2.39	.341
ALL-AF	1,148,956	264,067	.230
ALL-AC	186,568	15,280	.082
ALL-AF/AC	6.21	1.60	.258
Price (1967=100)			
WMC	98.9	12.5	.126
WMUS	101.9	10.3	.101
WHAY	103.1	13.8	.134
WFC	86.7	15.1	.174
WALLC	100.7	10.9	.108
Net Revenue			
WREV	184.0	83.0	.451
Precipitation			
JUNC	7.8	2.1	.269
RIFLE	10.6	2.3	.217

SOURCES: Division Five Engineer's annual reports, 1960-1978; *Colorado Agricultural Statistics*; and *U.S. Climatological Data*, annual summaries.

WMC	=	deflated index of Colorado livestock prices
WMUS	=	deflated index of U.S. livestock prices
WHAY	=	deflated index of Colorado hay and forage prices
WFC	=	deflated index of Colorado fruit prices
WALLC	=	deflated index of all Colorado agricultural product prices
WREV	=	deflated value of Net Income Realized in Colorado agriculture (Millions of dollars; 1967 base year)
JUNC	=	annual precipitation at Grank Junction, Colorado (inches)
RIFLE	=	annual precipitation at Rifle, Colorado, (inches)
ALL-AF	=	water diversion rates from all four water districts (39, 45, 70, and 72)
ALL-AC	=	irrigated acreage in all water districts
ALL-AF/AC	=	irrigation application rates in all water districts

TABLE 41

TRENDS IN WATER DIVERSION AND INDEPENDENT VARIABLES, 1960-1978
(n = 18)

VARIABLE	TREND	AUTOCORR COEF (lag 1)	ADJ. BOX JENKINS (lag 1)
Water Diversion			
39 AF	-.2418	.052	.60
39 AC	-.6743**	.572	7.24
39 AF/AC	.3797	.382	3.24
45 AF	-.2584	-.328	2.38
45 AC	.0148	.057	.07
45 AF/AC	-.2783	.307	2.09
70 AF	.7091**	.404	3.62
70 AC	.4510+	.265	1.55
70 AF/AC	.6249**	.306	2.08
72 AF	.5290*	.310	2.13
72 AC	.1755	.169	.64
72 AF/AC	.4043+	.136	.41
ALL-AF	.5221*	.292	1.89
ALL-AC	.1026	.191	.81
ALL-AF/AC	.4571+	.138	.42
Price Variables			
WMC	.2083	.568	7.15
WMUS	.2579	.387	3.33
WHAY	.3499	.631	8.82
WFC	.0490	.479	5.08
WALLC	.2461	.638	9.04
Net Revenue			
WREV	.3890	.734	11.94
Precipitation			
JUNC	-.1546	-.133	.39
RIFLE	.3912	-.027	.02

+α = .10; *α = .05; **α = .01; d.f. = n-2; underlining indicates that the auto-correlation coef. exceeds two standard error limits.

TABLE 42

CORRELATION OF WATER USE VARIABLES WITH PRICE, INCOME,
AND PRECIPITATION VARIABLES

District Variable	WMC	WMUS	WHAY	WFC	WALLC	WREV	JUNC	RIFLE	n
39 AF	.0935	.0740	-.1802	.0429	.0980	.0523	-.1741	-.2349	18
39 AC	.0742	-.1525	-.5125*	.1189	-.1473	-.3478	.1452	.0877	18
39 AF/AC	.0722	.2169	.3037	-.0059	.2201	.3279	-.2054	-.1758	18
45 AF	.1484	.1583	.0824	.0182	.1816	.1319	.2751	.0975	18
45 AC	-.0603	-.0571	-.4401+	--	-.2664	-.2604	.0395	.3552	17
45 AF/AC	.3118	.3084	.4545+	--	.5070*	.4118	.3435	-.0096	17
70 AF	.2367	.3790	.5210*	-.0397	-.4140+	.5095*	.1111	.1873	18
70 AC	.2345	.2710	.3066	.0207	.3495	.5033*	-.1136	-.0128	18
70 AF/AC	.1515	.2792	.4048+	-.0883	.2750	.2498	.1750	.2550	18
72 AF	.3894	.2452	.1924	.1776	.4110+	.5108*	-.0633	.2399	18
72 AC	-.1466	.0446	.5793*	.1049	.1360	.3893	-.0828	-.4122+	18
72 AF/AC	.4026+	.2084	-.0963	.0963	.2908	.2576	-.0259	.3905	18
ALL-AF	.4051+	.2838	.2271	.1666	.4454+	.5410*	-.0427	.2287	18
ALL-AC	-.1086	.0583	.3806	--	.0721	.3053	-.0152	-.2913	17
ALL-AF/AC	.4422+	.2800	.0591	--	.3922	.3793	-.0218	.3871	17

+α = .10; *α = .05; **α = .01; d.f. = n-2; two-tailed t test

TABLE 43

CORRELATION RESULTS FOR ANNUAL DIVERSION RECORDS (LAG 1)

District Variable	WMC	WMUS	WHAY	WFC	WALLC	WREV	JUNC	RIFLE	n
39 AF	.2177	.1936	.2168	.0986	.1384	.1769	-.0859	-.0849	18
39 AC	.0613	-.1738	-.5494*	.2159	-.2565	-.4733*	.1141	.0181	18
39 AF/AC	.2310	.2884	.2485	-.2220	.3061	.4937*	-.1793	.1113	18
45 AF	.1219	-.2126	-.0912	.0399	-.1163	.0115	-.0389	-.2468	18
45 AC	-.5380*	-.5836*	-.4190+	-.4096	-.6095**	-.3228	-.5401*	-.0879	17
45 AF/AC	.3893*	.4280+	.2290	.5274*	.5225*	.3987	.4532+	-.1203	17
70 AF	.2317	.2872	.3424	.2761	.3283	.4150*	.1251	.3833	18
70 AC	.3853	.4218+	.1801	.1117	.4263+	.5104*	-.0838	.2811	18
70 AF/AC	.0781	.1476	.3961	.2971	.1875	.2575	.1895	.3527	18
72 AF	.4977*	.4836*	.0976	.1056	.4414+	.3906	.0206	.4687*	18
72 AC	.2398	.2590	.6406**	.3521	.4739*	.6592**	.0729	-.3229+	18
72 AF/AC	.2903	.2743	-.1817	-.0642	.1462	-.0412	-.0152	.5551*	18
ALL-AF	.4926*	.4770*	.1023	.1367	.4447+	.4187+	-.0742	.4623+	18
ALL-AC	.1118	.0908	.4657+	.2508	.3057	.5871*	-.2707	-.3711	17
ALL-AF/AC	.3963	.3851	-.0669	.0591	.2835	.1801	.0294	.5623*	17

+α = .10; *α = .05; **α = .01; d.f. = n-2; two-tailed t test

TABLE 44

LAND USE SUMMARY FOR THE GRAND VALLEY AREA,
BY CANAL SERVICE AREA, 1969 (acres)

	GRAND VALLEY PROJECT CANALS			Orchard Mesa #1	Orchard Mesa #2	MUTUAL COMPANY CANALS			TOTAL
	Govt. Highline	Stub Ditch	Price Ditch			Grand Valley	Mesa County	Redlands Canals	
Total Irrigated	25,169	608	4,306	4,665	3,029	28,407	1,320	3,406	70,550
Cropland	15,414	215	1,469	1,811	468	16,892	572	859	37,700
Forage & Pasture	9,060	146	1,262	1,202	720	11,001	707	1,816	25,914
Orchard	695	247	1,575	1,652	1,841	514	41	371	6,936
Urban Uses	1,629	32	547	1,178	–	5,421	119	885	9,996
Industrial	–	–	–	–	–	661	21	–	682
Open Water	635	31	37	88	47	750	48	63	1,699
Phreatophytes	6,554	51	177	483	367	5,995	345	1,202	15,174
Precip. Surfaces	10,429	51	337	462	502	3,540	51	1,235	16,607
TOTAL	44,416	773	5,404	6,876	4,130	44,774	1,904	6,431	114,708

SOURCE: Wynn R. Walker and Gaylord V. Skogerboe, *Agricultural Land Use in the Grand Valley* (Fort Collins Co: Colorado State University, 1971), table 79.

TABLE 45

JOINT FREQUENCY DISTRIBUTION OF FIELD SIZE (ACRES)
BY CANAL SERVICE DISTRICT

Canal	0-10	10-20	20-30	30-40	40-50	50-60	Total*
Public Canals - West							
Gov't Highline	77	34	12	1	--	--	124
Private Canals - West							
Grand Valley Highline	21	2	2	2	1	--	28
Grand Valley Mainline	25	6	1	--	--	--	32
Kiefer Extension	19	14	3	2	1	1	40
Independent Ranchman's	3	1	--	1	--	--	5
Public Canals - East							
Stub Ditch	8	--	--	--	--	--	8
Price Ditch	33	2	--	--	--	--	35
Private Canals - East							
Grand Valley - East	21	1	--	--	--	--	22
Mesa County Ditch	3	--	--	--	--	--	3
Public Canals - South							
O.M. #1 (low)	16	3	1	--	--	--	20
O.M. #2 (high)	26	3	--	--	--	--	29
O.M. Power Canal	3	1	--	--	--	--	4
Private Canals - South							
Redlands Canal	12	2	--	--	--	--	14
TOTAL	170	69	20	7	2	1	269

SOURCE: U.S. Soil Conservation Service, unpublished field data.

*Data on field size was not available for 102 fields of this sample.

TABLE 46

FREQUENCY DISTRIBUTION OF EXISTING FLOW MEASUREMENT
SYSTEMS BY CANAL SERVICE DISTRICT

Canal	Not Present	Present	Total Fields	% Present
Public Canals - West				
Gov't Highline	89	35	124	28.2
Private Canals - West				
Grand Valley - Highline	28	1	29	3.4
Grand Valley - Mainline	32	-	32	-
Kiefer Ext.	39	1	40	2.5
Independent Ranchman's	5	-	5	-
Public Canals - East				
Stub Ditch	8	-	8	-
Price Ditch	35	-	35	-
Private Canals - East				
Grand Valley - East	21	1	22	4.5
Mesa County Ditch	3	-	3	-
Public Canals - South				
O.M. #1 (low)	18	2	20	10.0
O.M. #2 (high)	22	7	29	24.1
O.M. Power Canal	4	-	4	-
Private Canals - South				
Redlands Canals	14	1	15	6.7
TOTAL	318	48	366	13.1

TABLE 47

FREQUENCY DISTRIBUTION OF EXISTING PIPELINE SYSTEMS
BY CANAL SERVICE DISTRICT

Canal	Not Present	Present	Total Fields	%Present
Public Canals - West				
Gov't Highline	120	4	124	3.2
Private Canals - West				
Grand Valley - Highline	29	-	29	-
Grand Valley - Mainline	32	-	32	-
Kiefer Ext.	39	1	40	2.5
Independent Ranchman's	5	-	5	-
Public Canals - East				
Stub Ditch	8	-	8	-
Price Ditch	32	3	35	8.6
Private Canals - East				
Grand Valley - East	20	2	22	9.1
Mesa County Ditch	3	-	3	-
Public Canals - South				
O.M. #1 (low)	17	3	20	15.0
O.M. #2 (high)	22	7	29	24.1
O.M. Power Canal	3	1	4	25.0
Private Canals - South				
Redlands Canals	11	4	15	26.7
TOTAL	341	25	366	6.8%

TABLE 48

FREQUENCY DISTRIBUTION OF EXISTING CONCRETE DITCH LINING SYSTEMS BY CANAL SERVICE DISTRICT

Canal	Not Present	Present	Total Fields	%Present
Public Canals - West				
Gov't Highline	94	30	124	24.2
Private Canals - West				
Grand Valley - Highline	23	6	29	20.7
Grand Valley - Mainline	25	7	32	21.9
Kiefer Ext.	30	10	40	25.0
Independent Ranchman's	4	1	5	20.0
Public Canals - East				
Stub Ditch	8	–	8	–
Price Ditch	29	6	35	17.1
Private Canals - East				
Grand Valley - East	18	4	22	18.2
Mesa County Ditch	3	–	3	–
Public Canals - South				
O.M. #1 (low)	18	2	20	10.0
O.M. #2 (high)	27	2	29	6.9
O.M. Power Canal	4	–	4	–
Private Canals - South				
Redlands Canals	15	–	15	–
TOTAL	298	68	366	18.5

TABLE 49

FREQUENCY DISTRIBUTION OF EXPRESSED FLOW MEASUREMENT NEEDS
BY CANAL SERVICE DISTRICT

Canal	Not Present	Present	Total Fields	%Present
Public Canals - West				
Gov't Highline	85	39	124	31.5
Private Canals - West				
Grand Valley - Highline	18	11	29	37.9
Grand Valley - Mainline	11	21	32	65.6
Kiefer Ext.	24	16	40	40.0
Independent Ranchman's	4	1	5	20.0
Public Canals - East				
Stub Ditch	3	5	8	62.5
Price Ditch	17	18	35	51.4
Private Canals East				
Grand Valley - East	15	7	22	31.8
Mesa County Ditch	2	1	3	33.3
Public Canals - South				
O.M. #1 (low)	14	6	20	30.0
O.M. #2 (high)	18	11	29	37.9
O.M. Power Canal	1	3	4	75.0
Private Canals - South				
Redlands	10	5	15	33.3
TOTAL	222	144	366	39.3

TABLE 50

FREQUENCY DISTRIBUTION OF EXPRESSED PIPELINE NEEDS BY CANAL SERVICE DISTRICT

Canal	Not Present	Present	Total Fields	%Present
Public Canals - West				
Gov't Highline	112	12	124	9.7
Private Canals - West				
Grand Valley - Highline	24	5	29	17.2
Grand Valley - Mainline	29	3	32	9.4
Kiefer Ext	36	4	40	10.0
Independent Ranchman's	5	–	5	–
Public Canals - East				
Stub Ditch	6	2	8	25.0
Price Ditch	30	5	35	14.3
Private Canals - East				
Grand Valley - East	19	3	22	13.6
Mesa County - Ditch	2	1	3	33.3
Public Canals - South				
O.M. #1 (low)	15	5	20	25.0
O.M. #2 (high)	26	3	29	10.3
O.M. Power	3	1	4	25.0
Private Canals - South				
Redlands Canals	13	2	15	13.0
TOTAL	320	46	366	12.6

TABLE 51

FREQUENCY OF EXPRESSED DITCH LINING NEEDS
BY CANAL SERVICE DISTRICT

Canal	Not Present	Present	Total Fields	%Present
Public Canals - West				
Gov't Highline	59	65	124	52.4
Private Canals West				
Grand Valley - Highline	18	11	29	37.9
Grand Valley - Mainline	13	19	32	59.4
Kiefer Ext.	15	25	40	62.5
Independent Ranchman's	5	–	5	–
Public Canals - East				
Stub Ditch	2	6	8	75.0
Price Ditch	14	21	35	60.0
Private Canals - East				
Grand Valley - East	7	15	22	68.2
Mesa County - Ditch	3	–	3	–
Public Canals - South				
O.M. #1 (low)	7	13	20	65.0
O.M. #2 (high)	13	16	29	55.2
O.M. Power Canal	4	–	4	–
Private Canals - South				
Redlands Canals	10	5	15	33.3
TOTAL	170	196	366	53.6

TABLE 52

FREQUENCY DISTRIBUTION OF EXPRESSED LAND LEVELING NEEDS
BY CANAL SERVICE DISTRICT

Canal	Not Present	Present	Total Fields	%Present
Public Canals - West				
Gov't Highline	64	60	124	48.4
Private Canals - West				
Grand Valley - Highline	17	12	29	41.4
Grand Valley - Mainline	17	15	32	46.9
Kiefer Ext.	20	20	40	50.0
Independent Ranchman's	5	–	5	–
Public Canals - East				
Stub Ditch	4	4	8	50.0
Price Ditch	27	8	35	22.9
Private Canals - East				
Grand Valley - East	14	8	22	36.4
Mesa County Ditch	3	–	3	–
Public Canals - South				
O.M. #1 (low)	6	14	20	70.0
O.M. #2 (high)	18	11	29	37.9
O.M. Power Canals	1	3	4	75.0
Private Canals - South				
Redlands Canals	10	5	15	33.3
TOTAL	206	160	366	43.7

TABLE 53

FREQUENCY DISTRIBUTION OF EXPECTED FUTURE LAND USE BY CANAL SERVICE DISTRICT

Canal	Cropland	Orchard	Urban	Watershed	Total	%Urban
Public Canals - West						
Gov't Highline	103	3	11	6	123	8.9
Private Canals - West						
Grand Valley - Highline	21	-	5	1	27	18.5
Grand Valley - Mainline	27	1	3	1	32	9.4
Kiefer Ext. Independent	35	-	3	1	39	7.7
Ranchman's	4	-	1	--	5	20.0
Public Canals - East						
Stub Ditch	8	-	--	-	8	-
Price Ditch	18	8	5	2	33	15.2
Private Canals - East						
Grand Valley - East	18	3	1	--	22	4.5
Mesa County Ditch	1	-	1	--	2	50.0
Public Canals - South						
O.M. #1 (low)	13	6	1	-	20	5.0
O.M. #2 (high)	9	16	2	2	29	-
O.M. Power Canal	2	2	-	-	4	6.9
Private Canals - South						
Redlands Canals	9	1	5	--	15	33.3
TOTAL	268	40	38	13	359	10.6

TABLE 54

SALINITY CONTROL COST-SHARING ASSISTANCE BY CANAL
SERVICE DISTRICT, 1980

Canal	No. Farms	Amt. C-S	%Total C-S	C-S/Farm
Public Canals - West Gov't Highline	14	$ 83,473	20.5%	$5,962.36
Private Canals - West Grand Valley - West Kiefer Ext. Independent Ranchman's	16	$116,631	28.7%	$7,289.44
Public Canals - East Stub Ditch Price Ditch	8	$ 44,267	10.9%	$5,533.38
Private Canals - East Grand Valley - East Mesa County - Ditch	5	$ 30,228	7.4%	$6,045.60
Public Canals - South O.M. #1 (low) O.M. #2 (high) O.M. Power Canal	27	$132,292	32.5%	$4,899.70
TOTAL	70	$406,891	100.0%	$5,812.73

C-S = Cost-Sharing assistance in dollars.

TABLE 55

FREQUENCY DISTRIBUTION OF ACTUAL CONSERVATION EXPENDITURES
(NO. AND AMOUNT) BY CANAL SERVICE DISTRICT, 1980

Canal	Pipeline	Gated Pipe	Land Lev.	Lining	Total*
Public Canals - West					
No. of projects	8	4	2	3	14
Amount Spent	$ 62,049	$9,494	$9,743	$20,679	$105,348
% Total no.	57.1	28.6	14.3	21.4	--
% Total spent	58.9	9.0	9.2	19.6	--
Private Canals - West					
No. of projects	11	3	4	4	16
Amount spent	$ 95,696	$6,940	$9,035	$21,116	$136,820
% Total no.	68.8	18.8	25.0	25.0	--
% Total spent	69.9	5.1	6.6	15.4	--
Public Canals - East					
No. of projects	7	1	--	2	8
Amount spent	$ 40,441	$ 871	--	$ 9,544	$ 50,857
% Total no.	87.5	12.5	--	25.0	--
% Total spent	79.5	1.7	--	18.8	--
Private Canals - East					
No. of projects	3	--	1	1	5
Amount Spent	$ 20,536	--	$17,494	$ 3,045	$ 41,175
% Total no.	60.0	--	20.0	20.0	--
% Total spent	49.9	--	42.5	7.4	--
Public Canals - South					
No. of projects	22	14	1	--	27
Amount spent	$117,003	$23,377	$3,315	--	$161,302
% Total no.	81.5	51.9	3.7	--	--
% Total spent	72.5	14.5	2.1	--	--
TOTAL					
No. of projects	51	22	8	10	70
Amount spent	$355,725	$406,682	$39,587	$54,384	$495,502
% Total no.	72.9	31.4	11.4	14.3	--
% Total spent	67.8	8.2	8.0	11.0	--

*Totals do not necessarily sum due to additional miscellaneous expenditures and the occurrence of multiple improvements on individual projects.

BIBLIOGRAPHY

Books and Monographs

Ackerman, Edward A., and Lof, George O. *Technology in American Water Development*. Baltimore: Johns Hopkins University Press, 1959.

Anderson, Jay C., and Kleinman, Alan P. *Salinity Management Options for the Colorado River*. Logan, UT: Utah Water Research Laboratory, 1978.

Bagley, Jay M.; Kimball, Kirk R.; and Kapaloski, Lee. *Feasibility Study of Establishing a Water Rights Banking/Brokering Service in Utah*. Logan, UT: Utah Water Research Laboratory, 1980.

Beyer, Jacquelyn. *Integration of Grazing and Crop Agriculture: Resources Management Problems in the Uncompahgre Valley Irrigation Project*. Research Papers, no. 57. Chicago: University of Chicago, Department of Geography, 1957.

Boris, Constance, and Krutilla, John V. *Water Rights and Energy Development in the Yellowstone River Basin: An Integrated Analysis*. Baltimore: Johns Hopkins University Press for Resources for the Future, Inc., 1980.

Bos, Marinus G., and Nugteren, J. *On Irrigation Efficiencies*. Publication 19. Waginingen: International Institute for Land Reclamation and Improvement, 1974.

Bowden, Leonard W. *Diffusion of the Decision to Irrigate: Simulation of the Spread of a New Resource Management Practice in the Colorado Northern High Plains*. Research Papers, no. 97. Chicago: University of Chicago, Department of Geography, 1965.

Bradley, Raymond S. *Precipitation History of the Rocky Mountain States*. Boulder, CO: Westview Press, 1976.

Chalmers, John R. *Southwestern Groundwater Law*. Arid lands Resource Information Paper, no. 4. Tucson: Office of Arid Lands Studies, 1974.

Chorley, Richard J., ed. *Water, Earth and Man*. London: Methuen, 1969.

Coward, E. Walter, Jr. *Research Methodology in the Study of Irrigation Organizations: A Review of Approaches and Applications*. Seminar Report, no. 18. New York: Agricultural Development Council, 1978.

Evans, Robert G., et al. *Implementation of Agricultural Salinity Control Technology in Grand Valley*. EPA 600/2-78-160. Ada, OK: U.S. Environmental Protection Agency, 1978.

Evans, Robert G., et al. *Evaluation of Irrigation Methods for Salinity Control in Grand Valley*. EPA 600/2-78-161. Ada, OK: U.S. Environmental Protection Agency, 1978.

Evans, Robert G.; Walker, Wynn R.; and Skogerboe, Gaylord V. *Optimizing Salinity Control Strategies for the Upper Colorado River Basin.* Fort Collins, CO: Colorado State University, Department of Agricultural and Chemical Engineering, 1981.

Fradkin, Philip L. *A River No More: The Colorado River and the West.* New York: Alfred A. Knopf, 1981.

Frank, Michael D., and Beattie, Bruce R. *The Economic Value of Irrigation Water in the Western United States: An Application of Ridge Regression.* Texas Water Resources Institute Report, no. 99. College Station, TX: Texas A and M University, 1979.

Gray, S. Lee, and McKean, John R. *An Economic Analysis of Water Use in Colorado's Economy.* Environmental Resources Center Report, no. 70. Fort Collins, CO: Colorado State University, 1975.

Hirshleifer, Jack; DeHaven, James C.; and Milliman, Jerome W. *Water Supply: Economics, Technology and Policy.* Chicago: University of Chicago Press, 1960.

Howe, Charles W., and Easter, William K. *Interbasin Transfer of Water.* Baltimore: Johns Hopkins University Press, 1971.

Hudson, James. *Irrigation Water Use in the Utah Valley, Utah.* Research Papers, no. 79. Chicago: University of Chicago, Department of Geography, 1963.

Huffman, Roy E. *Irrigation and Public Water Policy.* New York: Ronald Press, 1953.

Hundley, Norris, Jr. *Dividing the Waters.* Berkeley: University of California Press, 1966.

_____. *Water and the West.* Berkeley: University of California Press, 1975.

Hutchins, Wells A. *Water Rights Laws in the Nineteen Western States.* vol. 2. U.S. Department of Agriculture, Miscellaneous Publication no. 1206. Washington, D.C., 1974.

Israelson, C. Earl, et al. *Use of Saline Water in Energy Development.* Water Resources Planning Series, no. UWRL/P-80/04. Logan, UT: Utah Water Research Laboratory, 1980.

Kasperson, Roger E., and Kasperson, Jeanne X., eds. *Water Reuse and the Cities.* Hanover, NH: University of New England Press, 1977.

Kelso, Maurice M.; Martin, William E.; and Mack, Laurence E. *Water Supplies and Economic Growth in an Arid Environment.* Tucson: University of Arizona Press, 1973.

Kinkaid, Del Breese. *Irrigation Law of Colorado.* Denver: W.H. Courtright and Co., 1912.

Maass, Arthur, and Anderson, Raymond L. *. . . and the Desert Shall Rejoice: Conflict, Growth, and Justice in Arid Environments.* Cambridge, MA: MIT Press, 1978.

Mead, Elwood. *Irrigation Institutions.* New York: Macmillan Co., 1903.

Milliken, J. Gordon, et al. *State and Local Management Actions to Reduce Colorado River Salinity.* For the U.S. Environmental Protection Agency. EPA 908/3-77-002. Washington, D.C.: Government Printing Office, 1978.

_____. *Overcoming Legal and Institutional Barriers to Planned Reuse of Water in the Colorado River Basin*. OWRT.RU-79/3. Washington, D.C.: U.S. Department of the Interior, Office of Water Research and Technology, 1979.

Morris, Glen E. *The Economic Impact of Synfuels Development in the Upper Colorado River Basin*. LA-UR-2784. Los Alamos, NM: Los Alamos Scientific Laboratory, 1979.

Mulder, Jim, et al. *Integrated Water Resources and Land Use Planning*. Logan, UT: Utah Water Research Laboratory, 1979.

National Academy of Sciences. *Water and Choice in the Colorado Basin: An Example of Alternatives in Water Management*. Washington, D.C.: National Academy of Sciences, 1968.

Peterson, Dean F.; and Crawford, A. Berry, eds. *Values and Choices in the Development of the Colorado River Basin*. Tucson: University of Arizona Press, 1977.

Probstein, Ronald F., and Gold, Harris. *Water in Synthetic Fuel Production: The Technology and Alternatives*. Cambridge: MIT Press, 1978.

Radosevich, George E.; Hamburg, Donald H.; and Swick, Loren L., eds. *Colorado Water Laws*. Fort Collins, CO: Colorado State University, 1975.

Radosevich, George E. *Western Water Laws and Irrigation Return Flow*. Ada, OK: U.S. Environmental Protection Agency, 1978.

Radosevich, George E.; and Skogerboe, Gaylord V. *Achieving Irrigation Return Flow Quality Control Through Improved Legal Systems*. EPA 600/2-78-184. Ada, OK: U.S. Environmental Protection Agency, 1978.

Ruttan, Vernon W. *The Economic Demand for Irrigated Acreage*. Baltimore: Johns Hopkins University Press for Resources for the Future, Inc., 1965.

Schurr, Sam H., et al. *Energy in America's Future*. Baltimore: Johns Hopkins University Press for Resources for the Future, Inc., 1979.

Skogerboe, Gaylord V., and Walker, Wynn R. *Evaluation of Canal Lining for Salinity Control in Grand Valley*. EPA-R2-72-047. Washington, D.C.: U.S. Environmental Protection Agency, 1972.

Skogerboe, Gaylord V., et al. *Evaluation of Irrigation Scheduling for Salinity Control in Grand Valley*. EPA-660/2-74-052. Washington, D.C.: U.S. Environmental Protection Agency, 1974.

_____. *Evaluation of Drainage for Salinity Control*. EPA-660/2-74-084. Washington, D.C.: U.S. Environmental Protection Agency, 1974.

_____., et al. *Potential Effects of Irrigation Practices on Crop Yields in Grand Valley*. EPA-600/2-79-149. Ada, OK: U.S. Environmental Protection Agency, 1979.

Skogerboe, Gaylord V.; McWhorter, David B.; and Ayars, James E. *Irrigation Practices and Return Flow Salinity in Grand Valley*. EPA-600/2-79-148. Ada, OK: U.S. Environmental Protection Agency, 1979.

Skogerboe, Gaylord V.; Walker, Wynn R.; and Evans, Robert G. *Environmental Planning Manual for Salinity Management in Irrigated Agriculture*. EPA-600/2-79-062. Ada, OK: U.S. Environmental Protection Agency, 1979.

United Nations. Department of Economic and Social Affairs. *Integrated River Basin Development*. New York, 1958.

University of Colorado. Bureau of Business Research. *Report on the Economic Potential of Western Colorado*. Denver, CO: Colorado Water Conservation Board, 1953.

University of Oklahoma. *Energy from the West: Policy Analysis Report*. EPA-600-7-79-083. Washington, D.C.: Government Printing Office, 1979.

University of Wisconsin. Water Resources Management Workshop. *Oil Shale Development in Northwestern Colorado: Water and Related Land Impacts*. Madison, WI: Institute for Environmental Studies, 1975.

Vandenbusche, Duane, and Smith, Duane A. *A Land Alone: Colorado's Western Slope*. Boulder, CO: Pruett Press, 1981.

Vlachos, Evan C., et al. *Consolidation of Irrigation Systems Phase II: Engineering, Economics, Legal and Sociological Requirements*. Completion Report no. 94. Fort Collins, CO: Colorado State University, 1980.

Walker, Wynn R., and Skogerboe, Gaylord V. *Agricultural Land Use in the Grand Valley*. Fort Collins, CO: Colorado State University, Agricultural Engineering Department, 1971.

Walker, Wynn R. *Integrating Desalination and Agricultural Salinity Control Measures*. EPA-600/2-78-074. Ada, OK: U.S. Environmental Protection Agency, 1978.

Walker, Wynn R.; Skogerboe, Gaylord V.; and Evans, Robert G. *"Best Management Practices" for Salinity Control in Grand Valley*. EPA-600/2-78-162. Ada, OK: U.S. Environmental Protection Agency, 1978.

Waters, Frank. *The Colorado*. New York: Holt, Rinehart, Winston, 1946.

White, Gilbert F. *Strategies of American Water Management*. Ann Arbor, MI: University of Michigan Press, 1969.

Whittlesey, Norman K. *Irrigation Development Potential in Colorado*. Report to the Colorado Department of Natural Resources, November 1977.

Wollman, Nathaniel. *The Value of Water in Alternative Uses*. Albuquerque, NM: University of New Mexico Press, 1962.

Articles

Abbey, David. "Energy Production and Water Resources in the Colorado River Basin." *Natural Resources Journal* 19 (1979): 275-314.

Anderson, Jay C., and Keith, John E. "Energy and the Colorado River." *Natural Resources Journal* 17 (1977): 157-168.

Anderson, Raymond L. "The Irrigation Water Rental Market: A Case Study." *Agricultural Economics Research* 13 (1961): 54-58.

_____. "Organizational Arrangements in Water Transfers." *Water Resources and the Economic Development of the West* 10 (1961): 1-8.

Anderson, Raymond L., and Wengert, Norman I. "Developing Competition for Water in the Urbanizing Areas of Colorado." *Water Resources Bulletin* 13 (1977): 769-73.

Anderson, Terry L., and Hill, P.J. "The Evolution of Property Rights: A Study of the American West." *Journal of Law and Economics* 10 (1975): 163-179.

Angelides, Sortirus, and Bardach, Eugene. "Water Banking: How to Stop Wasting Agricultural Water." San Francisco: Institute for Contemporary Studies, 1978.

Beise, Charles J. "When Corporate Stock Becomes Real Estate." *Dicta* 21 (1944): 53-61.

Bergholz, Warren E., Jr. "Water Saved or Water Lost: The Consequences of Individual Conservation Measures in the Appropriation States." *Land and Water Law Review* 11 (1976): 435-457.

Bishop, A. Bruce, and Narayanan, Rangesan. "Competition of Energy for Agricultural Water Use." *Journal of the Irrigation and Drainage Division, ASCE* 105 (1979): 317-35.

Brown, F. Lee; Sawyer, James W.; and Khoshakhlagh, Rahman. "Some Remarks on Energy Related Water Issues in the Upper Colorado River Basin." *Natural Resources Journal* 17 (1977): 635-648.

Burness, H. Stuart, and Quirk, James P. "Appropriative Water Rights and the Efficient Allocation of Resources." *American Economic Review* 69 (1979): 25-37.

Carlson, John U. "Water Tabulation and Abandonment." In *Water Law 1978: Proceedings*, pp. 1-14. Denver: Colorado Bar Association, 1978.

Caulfield, Henry P., Jr. "Let's Dismantle (Largely but not Fully) the Federal Water Resource Development Establishment: The Apostasy of a Longstanding Water Development Federalist." In *Water Needs for the Future*, pp. 171-178. Edited by Ved P. Nanda. Boulder, CO: Westview Press, 1977.

Clyde, Edward W. "Current Problems in Water Acquisition--Legal Overview." In *Water Acquisition for Mineral Development Proceedings*, chap. 2. Boulder, CO: Rocky Mountain Mineral Law Foundation, 1978.

Corbridge, James N., Jr. "Outline: Application of the Law of Prior Appropriation." In *Proceedings: Water Resources Allocation, Laws and Emerging Issues*, chap. C. Boulder, CO: University of Colorado, School of Law, 1981.

Davenport, David C., and Hagan, Robert M. "A Conceptual Framework for Evaluating Agricultural Water Conservation Potential." *Water Resources Bulletin* 16 (1980): 717-723.

Davis, George H., and Kilpatrick, F.A. "Water Supply as a Limiting Factor in Western Energy Development." *Water Resources Bulletin* 17 (1981): 29-35.

Dempsey, Paul S. "Oil Shale and Water Quality: The Colorado Prospectus Under Federal, State and International Law." *Denver Law Journal* 58 (1981): 715-750.

Diemer, Joel A., and Wengert, Norman I. "Water for Energy: An Approach to Comprehensive Impact Assessment." *Water Resources Bulletin* 13 (1977): 885-893.

DuMars, Charles, and Ingram, Helen. "Congressional Quantification of Indian Reserved Water Rights." *Natural Resources Journal* 20 (1980): 17-43.

Dunning, Harrison C. "Reflections on the Transfer of Water Rights." *Journal of Contemporary Law* 4 (1977): 109-117.

El-Gasseir, Mohamed. "Coal Conversion and Shale Oil Extraction: The Environmental Implications for Water Resources and Aquatic Ecosystems." In *Energy and the Fate of Ecosystems,* pp. 122-263. Washington, D.C.: National Academy Press, 1981.

Ellis, Willis H., and DuMars, Charles. "The Two-Tiered Market in Western Water." *Nebraska Law Review* 57 (1978): 333-367.

Flug, Marshall; Walker, Wynn R.; and Skogerboe, Gaylord V. "Energy-Water-Salinity: Upper Colorado River Basin." *Journal of the Water Resources Planning and Management Division, ASCE* 105 (1979): 305-315.

Fox, Irving K. "Institutions for Water Management in a Changing World." *Natural Resources Journal* 16 (1976): 743-758.

Gallectus, Paul S. "The Transfer of Water Rights for Use in the Oil Industry." *Land and Water Law Review* 5 (1970): 441-448.

Gisser, Micha, et al. "Water Trade-Off Between Electric Energy and Agriculture in the Four Corners Area." *Water Resources Research* 15 (1979): 529-538.

Hampton, Norman F., and Ryan, Bennett Y., Jr. "Water Constraints on Emerging Energy Production." *Water Resource Bulletin* 16 (1980): 508-513.

Harrison, David L., and Sandstrom, Gustave, Jr. "The Groundwater--Surface Water Conflict and Recent Colorado Water Legislation." *University of Colorado Law Review* 43 (1971): 1-48.

Harrison, David L. "Federal Regulation of Appropriations of Water in the Name of Protecting Water Quality." In *Proceedings: Water Resources Allocation, Laws and Emerging Issues,* chap. P. Boulder, CO: University of Colorado, School of Law, 1981.

Hart, Gary. "Emerging Values in Water Resources Management." *Denver Journal of International Law and Policy* 6 (1976): 357-361.

Hart, William E., et al. "Irrigation Performance: An Evaluation," *Journal of the Irrigation and Drainage Division, ASCE* 105 (1979): 275-288.

Harte, John, and El-Gasseir, Mohamed. "Energy and Water." *Science* 199 (1978): 623-634.

Holburt, Myron B. "International Problems." In *Values and Choices in the Development of the Colorado River Basin,* pp. 220-238. Edited by Dean F. Peterson and A. Berry Crawford. Tucson: University of Arizona Press, 1977.

Hope, Mary. "Town of DeBeque v. Enewold: Conditional Water Rights and Statutory Water Law." *Denver Law Journal* 58 (1981): 837-846.

Howitt, Richard E.; Watson, William D.; and Adams, Richard M. "A Reevaluation of Price Elasticities for Irrigation Water." *Water Resources Research* 16 (1980): 623-628.

Kelly, William R. "Water Conservancy Districts." *Rocky Mountain Law Review* 22 (1950): 432-452.

_____. "Rehabilitation and Reorganization of Irrigation Projects that Parallel or Duplicate One Another: Legal Problems in Colorado." *Water Resources and the Economic Development of the West* 7 (1958): 1-12.

Kelso, Maurice M. "Competition for Water in an Expanding Economy." In *Water Resources and the Economic Development of the West* 16 (1967): 187-196.

Keys, John W. III. "Grand Valley Irrigation Return Flow Case Study." *Journal of the Irrigation and Drainage Division, ASCE* 107 (1981): 221-232.

Laitos, Jan G. "The Effect of Water Law on the Development of Oil Shale." *Denver Law Journal* 58 (1981): 751-783.

McIntire, Michael V. "The Disparity Between State Water Rights Records and Actual Water Use Patterns." *Land and Water Law Review* 5 (1970): 23-48.

Meyers, Charles J. "The Colorado River." *Stanford Law Review* 19 (1966): 1-75.

_____. "Outline--The Colorado River Compact: A Limit on Upper Basin Development." In *Proceedings: Water Resources Allocation, Laws and Emerging Issues*. Boulder, CO: University of Colorado, School of Law, 1981.

Moses, Raphael J. "Irrigation Corporations." *Rocky Mountain Law Review* 32 (1960): 527-533.

Narayanan, Rangesan, and Padunchai, Sumol. "Effects of Energy Development in the Upper Colorado River Basin on Irrigated Agriculture and Salinity." *Western Journal of Agricultural Economics* 4 (1979): 73-81.

Novak, Benjamin. "Legal Classification of Special District Corporate Forms in Colorado." *Denver Law Journal* 45 (1968): 347-380.

Ostrom, Vincent. "The Water Economy and Its Organization." *Natural Resources Journal* 2 (1962): 57-73.

Pascoe, D. Monte. "Outline of Remarks." In *Proceedings: Water Resources Allocation, Laws and Emerging Issues*. Boulder, CO: University of Colorado, School of Law, 1981. (and typewritten outline.)

Quinn, Frank. "Water Transfers: Must the American West be Won Again?" *Geographical Review* 58 (1968): 108-132.

Radosevich, George E.; Vlachos, Evan C.; and Skogerboe, Gaylord V. "Constraints in Water Management on Agricultural Lands." *Water Resources Bulletin* 9 (1973): 352-359.

Radosevich, George E., et al. "Salinity Management Alternatives for Oil Shale Water Supplies." *Natural Resources Journal* 17 (1977): 461-475.

Robbins, David W., and Hamburg, Donald H. "Water Right Regulation and Administration, the State Engineer and Obtaining Administrative Relief." In *Water 1978: Proceedings*. Denver: Colorado Bar Association, 1978.

Robinson, Charlotte, and Walta, Mary E. "Note, Water for Oil Shale: Framework for the Legal Issues." *Denver Law Journal* 58 (1981): 703-714.

Skogerboe, Gaylord V., et al. "Salinity Policy for Colorado River Basin." *Journal of the Hydraulics Division, ASCE* 101 (1975): 1067-1075.

_____. "Salinity Planning Control Framework." *Journal of the Water Resources Planning and Management Division, ASCE* 105 (1979): 329-341.

Spreen, William C. "A Determination of the Effect of Topography Upon Precipitation." *Transactions, American Geophysical Union* 28 (1947): 285-290.

Strahle, Ronald H. "Problems in the Rehabilitation of Irrigation Projects Through Reorganization and Merger." *Water Resources and the Economic Development of the West* 6 (1957): 95-102.

Tolley, George S., and Hastings, V.S. "Optimal Water Allocation: The North Platte River." *Quarterly Journal of Economics* 74 (1960): 279-295.

Trelease, Frank J., and Lee, Dellas W. "Priority and Progress--Case Studies in the Transfer of Water Rights." *Land and Water Law Review* 1 (1966): 1-76.

Trelease, Frank J. "The Model Water Code, the Wise Administrator, and the Goddam Bureaucrat." *Natural Resources Journal* 14 (1974): 207-229.

_____. "Federal Reserved Rights Since PLLRC." *Denver Law Journal* 54 (1977): 473-492.

_____. "Federal-State Problems in Packaging Water Rights." In *Water Acquisition for Mineral Development: Proceedings.* Boulder, CO: Rocky Mountain Mineral Law Foundation, 1978.

_____. "The Changing Water Market for Energy Production." *Journal of Contemporary Law* 5 (1978): 83-93.

White, Gilbert F. "A Perspective of River Basin Development." *Law and Contemporary Problems* 22 (1957): 157-188.

_____. "Contributions of Geographical Analysis to River Basin Development." In *Readings in Resource Management and Conservation,* pp. 375-394. Edited by Ian Burton and Robert W. Kates. Chicago: University of Chicago Press, 1965.

_____. "Role of Geography in Water Resources Management." In *Man and Water,* pp. 102-121. Edited by L. Douglas James. Lexington, KY: University of Kentucky Press, 1974.

Williams, Stephen F. "Optimizing Water Use: The Return Flow Issue." *University of Colorado Law Review* 44 (1973): 301-321.

Wolman, M. Gordon. "Selecting Alternatives in Water Resources Planning." *Natural Resources Journal* 16 (1976): 773-791.

Young, Robert A. "Economic Analysis of Federal Irrigation Policy: A Reappraisal." *Western Journal of Agricultural Economics* 3 (1978): 257-267.

Government Publications--Federal

Davis, George H., and Wood, Leonard A. *Water Demands for Expanding Energy Development.* U.S. Geological Survey Circular, no. 703. Washington, D.C.: Government Printing Office, 1974.

Duncan, Donald C., and Swanson, Vernon E. *Organic-Rich Shale of the United States and World Land Areas*. U.S. Geological Survey Circular, no. 523. Washington, D.C.: Government Printing Office, 1965.

Iorns, W.V.; Hembree, C.H.; and Oakland, G.L. *Water Resources of the Upper Colorado River Basin*. U.S. Geological Survey Professional Paper, no. 441. Washington, D.C.: Government Printing Office, 1965.

Kleinman, Alan P., and Brown, Bruce F. *Colorado River Salinity*. Denver: U.S. Department of the Interior, Bureau of Reclamation, 1980.

Lohman, Stanley W. *Geology and Artesian Water Supply--Grand Junction Area, Colorado*. U.S. Geological Survey Professional Paper, no. 451. Washington, D.C.: Government Printing Office, 1965.

Murray, C. Richard, and Reeves, E. Bodette. *Estimated Use of Water in the United States*. U.S. Geological Survey Circular, no. 765. Washington, D.C.: Government Printing Office, 1977.

Steele, Timothy D., and Hillier, D.E. *Assessment of Impacts of Proposed Coal Resource and Related Economic Development on Water Resurces, Yampa River Basin, Colorado and Wyoming--A Summary*. U.S. Geological Survey Circular, no. 839. Washington, D.C.: Government Printing Office, 1981.

Thomas, Harold E. *Effects of Drought in the Colorado River Basin*. U.S. Geological Survey Professional Paper, no. 372-F. Washington, D.C.: Government Printing Office, 1963.

U.S. Congress, Office of Technology Assessment. *An Assessment of Oil Shale Technologies*. Washington, D.C.: Government Printing Office, 1980.

U.S. Department of Agriculture. Agricultural Stabilization and Conservation Service. *National Summary Evaluation of the Agricultural Conservation Program, Phase I*. Washington, D.C.: Government Printing Office, 1980.

U.S. Department of Agriculture. Soil Conservation Service. *Irrigation Water Requirements*. Technical Release, no. 21. Washington, D.C.: Government Printing Office, 1967 (revised 1970).

_____. *On-Farm Program for Salinity Control: Final Report of the Grand Valley Salinity Study*. Washington, D.C.: Government Printing Office, 1977, and Supplement no. 1, 1980.

U.S. Department of Agriculture. Soil Conservation Service. Special Products Division. *Crop Consumptive Irrigation Requirements and Irrigation Efficiency Coefficients for the United States*. Portland, OR, 1976.

U.S. Department of Agriculture. Soil Conservation Service. Colorado State Office. *Colorado Irrigation Guide*. Denver, CO, n.d.

U.S. Department of the Interior. Bureau of Land Management. *Final Environmental Impact Statement, Proposed Development of Oil Shale Resources by the Colony Development Operation in Colorado*. 2 vols. Washington, D.C.: Government Printing Office, 1976.

_____. *West-Central Colorado Coal, Draft Environmental Statement*. 2 vols. Washington, D.C.: Government Printing Office, 1978.

_____. *Instream Flow Guidelines*. Denver, CO, 1979.

U.S. Department of the Interior. Bureau of Reclamation. *The Colorado River: A Natural Menace Becomes a National Resource*. H.R. Doc't no. 419, 80th Cong. 1st sess. Washington, D.C.: Government Printing Office, 1947.

_____. *Collbran Project, Colorado. Definite Plan Report*. Vol 1. Salt Lake City, UT, 1952.

_____. *Silt Project, Colorado. Definite Plan Report*. Salt Lake City, UT, 1961.

_____. "Colorado River Water Quality Improvement Program." Special Report, 1972.

_____. *Water for Tomorrow: Colorado State Water Plan. Phase Two, Legal and Institutional Considerations*. Denver, CO, 1974.

_____. *Stage One Development, Grand Valley Unit. Definite Plan Report*. n.p., 1980.

_____. *Colorado River Basin Salinity Control Project, Grand Valley Unit, Stage One Development: Monitoring Plan*. n.p., 1980.

_____. *Potential Modifications in Eight Proposed Western Colorado Projects for Future Energy Development: Special Report*. n.p., 1980.

_____. *Progress Report: Appraisal Investigation, Saline Water Use and Disposal Opportunities, Colorado River Water Quality Improvement Program*. Boulder City, NV, 1980.

_____. *Annual Operating Plan: Fryingpan-Arkansas Project Colorado. 1980 Operations, 1981 Outlook*. Denver, CO, 1980.

_____. *Quality of Water: Colorado River Basin. Progress Report no. 10*. Washington, D.C.: Government Printing Office, 1981.

_____. "Colorado-Big Thompson Project: Proposed Green Mountain Reservoir Operating Policy." *Federal Register* 46 (March 26, 1981): 18801-18802.

_____. "Saline Water Use and Disposal Opportunities. Draft Special Report." Denver, CO, 1981.

_____. "Draft: An Executive Summary of the Colorado River Simulation System." Denver, CO, 1981. (Typewritten.)

_____. *Final Environmental Statement: Colorado-Big Thompson Windy Gap Project, Colorado*. Denver, CO, 1981.

U.S. Department of the Interior. Bureau of Reclamation, and U.S. Department of Agriculture. Soil Conservation Service. *Final Environmental Statement: Colorado River Water Quality Improvement Program*. 2 vols. Washington, D.C.: Government Printing Office, 1977.

U.S. Department of the Interior. *Final Environmental Impact Statement for the Prototype Oil Shale Leasing Program*. 6 vols. Washington, D.C.: Government Printing Office, 1973.

U.S. Department of the Interior. Geological Survey. *Summary Appraisals of the Nation's Ground-Water Resources, Upper Colorado Region*. Professional Paper, no. 813-C. Washington, D.C.: Government Printing Office, 1974.

_____. *Availability and Chemical Quality of Groundwater in the Crystal River and Cattle Creek Drainage Basins Near Glenwood Springs, West-Central Colorado*. Water Resources Investigation Open-File Report, no. 76-70. Lakewood, CO, 1976.

U.S. Environmental Protection Agency. "The Mineral Quality Problem in the Colorado River Basin." In *Conference in the Matter of Pollution of the Interstate Waters of the Colorado River and its Tributaries, Seventh Session*, vol. 1, 1972.

_____. *Energy from the West, Volume II: Site Specific and Regional Impact Analysis*. Chap. 6, "The Rifle Area." Washington, D.C.: Government Printing Office, 1979.

_____. *Environmental Perspective on the Emerging Oil Shale Industry*. EPA-600/2-80-205a. Cincinnati, OH, 1980.

_____. *Final Environmental Impact Statement: Northglenn Water Management Program, City of Northglenn, Colorado*. Denver, CO, 1980.

U.S. General Accounting Office. *Report to the Congress, January 24, 1980--Water Supply Should not be an Obstacle to Meeting Energy Development Goals*. Washington, D.C.: Government Printing Office, 1980.

U.S. House of Representatives. *Operating Principles: Fryingpan Arkansas Project*. H. Doc. 130, 87th Cong., 1st sess., 1961.

U.S. Interagency Task Force. *Irrigation Water Use and Management*. Washington, D.C.: Government Printing Office, 1979.

U.S. Senate. *Synopsis of Report on the Colorado-Big Thompson Project, Plan of Development and Cost Estimate*. S. Doc. 80, 75th Cong., 1st sess., 1937.

U.S. Water Resources Council. *The Nation's Water Resources: 1975-2000*. Washington, D.C.: Government Printing Office, 1978.

_____. *Synthetic Fuels Development in the Upper Colorado Region: Section 13(a) Water Assessment Report*. Washington, D.C.: Government Printing Office, 1981.

Weeks, John B., et al. *Simulated Effects of Oil Shale Development on the Hydrology of Piceance Creek Basin, Colorado*. U.S. Geological Survey Professional Paper, no. 908. Washington, D.C.: Government Printing Office, 1974.

Government Publications--State and Local

City of Grand Junction, Department of Utilities. "A Report on the City of Grand Junction Water Supply System." Grand Junction, CO, 1977. (Typewritten.)

Colorado. Department of Agriculture. *Agricultural Land Conversion in Colorado*, 1980.

Colorado. Department of Natural Resources. *Annual Report of the Division 5 Engineer*, 1950-1979.

_____. *The Availability of Water for Oil Shale and Coal Gasification Development in the Upper Colorado River Basin*. Draft Summary Report to the U.S. Water Resources Council, 1979.

Colorado Energy Research Institute. *Water and Energy in Colorado's Future*. Boulder, CO: Westview Press, 1980.

Colorado. Water Conservation Board. *The Green Mountain Reservoir Problem,* 1960.

_____. *Water and Related Land Resources, Colorado River Basin in Colorado,* 1965.

_____. *Water and Related Land Resources, White River Basin in Colorado,* 1966.

Colorado. Water Quality Control Commission. "Colorado River Salinity Standards." Memorandum, 6 May 1980.

Colorado West Area Council of Governments. *208 Plan, Final Main Report,* 1979.

_____. *1981 Oil Shale Trust Fund Requests,* 1980.

Denver Regional Council of Governments. *Regional Water Study,* 1978.

Knudsen, Walter I., Jr. "Colorado Water Data Bank." Colorado Division of Water Resources, 1980. (Typewritten.)

Orchard Mesa Irrigation District. *Report on Rehabilitation and Betterment Program: Pumping and Conveyance Facilities.* Grand Junction, CO, n.d.

Metropolitan Denver Water Study Committee. *Metropolitan Water Requirements and Resources, 1975-2000,* 3 vols. Denver, CO, 1975.

Upper Colorado River Commission. *Thirty-Second Annual Report.* Salt Lake City, UT, 1978.

Unpublished Materials

Beale, Henry B.P. "The Use of Capitalized Values of Land and Water Rights in Evaluating Irrigation Projects." Ph.D. dissertation, Department of Economics, University of Chicago, 1974.

Burness, H. Stuart, and Quirk, James P. "Water Law, Water Transfers and Economic Efficiency." Social Science Working Paper, no. 28. Pasadena, CA: California Institute of Technology, 1978.

David E. Flemming Co. "The CORSIM Project." Paper presented at the annual meeting of the American Society of Civil Engineers, Denver, CO, 1975.

_____. "Feasibility Report: West Divide Project, South Rifle Division, First Stage." For the West Divide Water Conservancy District, 1980.

Gould, George A. "Conversion of Agricultural Water Rights to Industrial Use." Paper presented to the Rocky Mountain Mineral Law Institute in San Diego, CA, July 1981. (Proceedings in press.)

Holt, Kent. "Division 6--Water Budget Program." Steamboat Springs: Irrigation Division No. 6, n.d. (Typewritten.)

Lavender, David. "Water in the West: The Future is Now." 1979. (Mimeographed.)

_____. "Water and the Western Slope." 1979. (Mimeographed.)

Issacson, Morton S. "Progress Report: Water Consumption by Energy Resource Developments in the Upper Colorado River Basin--A Range of Estimates for the Year 2000." Pasadena, CA: California Institute of Technology, Environmental Quality Laboratory, 1980.

Lohman, Loretta C., and Milliken, J. Gordon. "Financial Incentives for the Electric Utilities to Reuse Low Quality Waters in the Colorado River Basin." Paper presented at the 1981 Water Reuse Symposium II in Washington, D.C. August 23-28, 1981.

Mehls, Steven F. "The Valley of Opportunity: A History of West Central Colorado." U.S. Department of the Interior, Bureau of Land Management, Colorado State Office (1980). (Typewritten draft.)

Mesa County, City/County Development Department. "Public Facilities in the Mesa County 201 Planning Area, 1981--Draft." Grand Junction, CO, 1981. (Typewritten.).

Miller, Robert W. "Cost-Sharing for Water Projects: Past, Present and Future." Paper presented to the Council of State Governments, Sacramento, CA, 12 June 1981.

Musick and Cope. "Colorado Water Law and Clean Water by Irrigation with Sewage Effluent." Boulder, CO, 1981. (Typewritten.)

_____. "An Introduction to Colorado Water Law." Boulder, CO, 1981. (Typewritten.)

_____. "Briefing Paper: Water for Western Energy Development." Boulder, CO, 1981. (Typewritten.)

_____. "Reclaiming Municipal Wastewater for Agricultural Use and Groundwater Considerations of Such Use." Boulder, CO, 1981. (Typewritten.)

Smith, Rodney T. "Private and Public Ownership of Agricultural Water Districts." Center for the Study of the Economy and the State, University of Chicago, 1980.

Ute Water Conservancy District. Memoranda on rates and policies. Grand Junction, CO, 1981. (Typewritten.)

Wright Water Engineers. "Water Rights and Water Resources Analysis for Battlemeht Mesa Inc. and Battlement Mesa Farm Lands, Garfield County, Colorado." Denver, CO, 1977.

_____. "Ground-Water Resources of Garfield County." Report to the Garfield County Planning Office, 1977.

THE UNIVERSITY OF CHICAGO
DEPARTMENT OF GEOGRAPHY
RESEARCH PAPERS (Lithographed, 6×9 inches)

LIST OF TITLES IN PRINT

48. BOXER, BARUCH. *Israeli Shipping and Foreign Trade*. 1957. 162 p.
56. MURPHY, FRANCIS C. *Regulating Flood-Plain Development*. 1958. 216 pp.
62. GINSBURG, NORTON, editor. *Essays on Geography and Economic Development*. 1960. 173 p.
71. GILBERT, EDMUND WILLIAM. *The University Town in England and West Germany*. 1961. 79 p.
72. BOXER, BARUCH. *Ocean Shipping in the Evolution of Hong Kong*. 1961. 108 p.
91. HILL, A. DAVID. *The Changing Landscape of a Mexican Municipio, Villa Las Rosas, Chiapas*. 1964. 121 p.
97. BOWDEN, LEONARD W. *Diffusion of the Decision To Irrigate: Simulation of the Spread of a New Resource Management Practice in the Colorado Northern High Plains*. 1965. 146 pp.
98. KATES, ROBERT W. *Industrial Flood Losses: Damage Estimation in the Lehigh Valley*. 1965. 76 pp.
101. RAY, D. MICHAEL. *Market Potential and Economic Shadow: A Quantitative Analysis of Industrial Location in Southern Ontario*. 1965. 164 p.
102. AHMAD, QAZI. *Indian Cities: Characteristics and Correlates*. 1965. 184 p.
103. BARNUM, H. GARDINER. *Market Centers and Hinterlands in Baden-Württemberg*. 1966. 172 p.
105. SEWELL, W. R. DERRICK, et al. *Human Dimensions of Weather Modification*. 1966. 423 p.
106. SAARINEN, THOMAS FREDERICK. *Perception of the Drought Hazard on the Great Plains*. 1966. 183 p.
107. SOLZMAN, DAVID M. *Waterway Industrial Sites: A Chicago Case Study*. 1967. 138 p.
108. KASPERSON, ROGER E. *The Dodecanese: Diversity and Unity in Island Politics*. 1967. 184 p.
109. LOWENTHAL, DAVID, editor, *Environmental Perception and Behavior*. 1967. 88 p.
112. BOURNE, LARRY S. *Private Redevelopment of the Central City, Spatial Processes of Structural Change in the City of Toronto*. 1967. 199 p.
113. BRUSH, JOHN E., and GAUTHIER, HOWARD L., JR., *Service Centers and Consumer Trips: Studies on the Philadelphia Metropolitan Fringe*. 1968. 182 p.
114. CLARKSON, JAMES D., *The Cultural Ecology of a Chinese Village: Cameron Highlands, Malaysia*. 1968. 174 p.
115. BURTON, IAN, KATES, ROBERT W., and SNEAD, RODMAN E. *The Human Ecology of Coastal Flood Hazard in Megalopolis*. 1968. 196 p.
117. WONG, SHUE TUCK, *Perception of Choice and Factors Affecting Industrial Water Supply Decisions in Northeastern Illinois*. 1968. 93 p.
118. JOHNSON, DOUGLAS L. *The Nature of Nomadism: A Comparative Study of Pastoral Migrations in Southwestern Asia and Northern Africa*. 1969. 200 p.
119. DIENES, LESLIE. *Locational Factors and Locational Developments in the Soviet Chemical Industry*. 1969. 262 p.
120. MIHELIČ, DUŠAN. *The Political Element in the Port Geography of Trieste*. 1969. 104 p.
121. BAUMANN, DUANE D. *The Recreational Use of Domestic Water Supply Reservoirs: Perception and Choice*. 1969. 125 p.
122. LIND, AULIS O. *Coastal Landforms of Cat Island, Bahamas: A Study of Holocene Accretionary Topography and Sea-Level Change*. 1969. 156 p.
123. WHITNEY, JOSEPH B. R. *China: Area, Administration and Nation Building*. 1970. 198 p.
124. EARICKSON, ROBERT. *The Spatial Behavior of Hospital Patients: A Behavioral Approach to Spatial Interaction in Metropolitan Chicago*. 1970. 138 p.
125. DAY, JOHN CHADWICK. *Managing the Lower Rio Grande: An Experience in International River Development*. 1970. 274 p.
126. MACIVER, IAN. *Urban Water Supply Alternatives: Perception and Choice in the Grand Basin Ontario*. 1970. 178 p.
127. GOHEEN, PETER G. *Victorian Toronto, 1850 to 1900: Pattern and Process of Growth*. 1970. 278 p.
128. GOOD, CHARLES M. *Rural Markets and Trade in East Africa*. 1970. 252 p.
129. MEYER, DAVID R. *Spatial Variation of Black Urban Households*. 1970. 127 p.
130. GLADFELTER, BRUCE G. *Meseta and Campiña Landforms in Central Spain: A Geomorphology of the Alto Henares Basin*. 1971. 204 p.

131. NEILS, ELAINE M. *Reservation to City: Indian Migration and Federal Relocation.* 1971. 198 p.
132. MOLINE, NORMAN T. *Mobility and the Small Town, 1900–1930.* 1971. 169 p.
133. SCHWIND, PAUL J. *Migration and Regional Development in the United States.* 1971. 170 p.
134. PYLE, GERALD F. *Heart Disease, Cancer and Stroke in Chicago: A Geographical Analysis with Facilities, Plans for 1980.* 1971. 292 p.
135. JOHNSON, JAMES F. *Renovated Waste Water: An Alternative Source of Municipal Water Supply in the United States.* 1971. 155 p.
136. BUTZER, KARL W. *Recent History of an Ethiopian Delta: The Omo River and the Level of Lake Rudolf.* 1971. 184 p.
139. MCMANIS, DOUGLAS R. *European Impressions of the New England Coast, 1497–1620.* 1972. 147 p.
140. COHEN, YEHOSHUA S. *Diffusion of an Innovation in an Urban System: The Spread of Planned Regional Shopping Centers in the United States, 1949–1968,* 1972. 136 p.
141. MITCHELL, NORA. *The Indian Hill-Station: Kodaikanal.* 1972. 199 p.
142. PLATT, RUTHERFORD H. *The Open Space Decision Process: Spatial Allocation of Costs and Benefits.* 1972. 189 p.
143. GOLANT, STEPHEN M. *The Residential Location and Spatial Behavior of the Elderly: A Canadian Example.* 1972. 226 p.
144. PANNELL, CLIFTON W. *T'ai-chung, T'ai-wan: Structure and Function.* 1973. 200 p.
145. LANKFORD, PHILIP M. *Regional Incomes in the United States, 1929–1967: Level, Distribution, Stability, and Growth.* 1972. 137 p.
146. FREEMAN, DONALD B. *International Trade, Migration, and Capital Flows: A Quantitative Analysis of Spatial Economic Interaction.* 1973. 201 p.
147. MYERS, SARAH K. *Language Shift Among Migrants to Lima, Peru.* 1973. 203 p.
148. JOHNSON, DOUGLAS L. *Jabal al-Akhdar, Cyrenaica: An Historical Geography of Settlement and Livelihood.* 1973. 240 p.
149. YEUNG, YUE-MAN. *National Development Policy and Urban Transformation in Singapore: A Study of Public Housing and the Marketing System.* 1973. 204 p.
150. HALL, FRED L. *Location Criteria for High Schools: Student Transportation and Racial Integration.* 1973. 156 p.
151. ROSENBERG, TERRY J. *Residence, Employment, and Mobility of Puerto Ricans in New York City.* 1974. 230 p.
152. MIKESELL, MARVIN W., editor. *Geographers Abroad: Essays on the Problems and Prospects of Research in Foreign Areas.* 1973. 296 p.
153. OSBORN, JAMES F. *Area, Development Policy, and the Middle City in Malaysia.* 1974. 291 p.
154. WACHT, WALTER F. *The Domestic Air Transportation Network of the United States.* 1974. 98 p.
155. BERRY, BRIAN J. L., et al. *Land Use, Urban Form and Environmental Quality.* 1974. 440 p.
156. MITCHELL, JAMES K. *Community Response to Coastal Erosion: Individual and Collective Adjustments to Hazard on the Atlantic Shore.* 1974. 209 p.
157. COOK, GILLIAN P. *Spatial Dynamics of Business Growth in the Witwatersrand.* 1975. 144 p.
159. PYLE, GERALD F. et al. *The Spatial Dynamics of Crime.* 1974. 221 p.
160. MEYER, JUDITH W. *Diffusion of an American Montessori Education.* 1975. 97 p.
161. SCHMID, JAMES A. *Urban Vegetation: A Review and Chicago Case Study.* 1975. 266 p.
162. LAMB, RICHARD F. *Metropolitan Impacts on Rural America.* 1975. 196 p.
163. FEDOR, THOMAS STANLEY. *Patterns of Urban Growth in the Russian Empire during the Nineteenth Century.* 1975. 245 p.
164. HARRIS, CHAUNCY D. *Guide to Geographical Bibliographies and Reference Works in Russian or on the Soviet Union.* 1975. 478 p.
165. JONES, DONALD W. *Migration and Urban Unemployment in Dualistic Economic Development.* 1975. 174 p.
166. BEDNARZ, ROBERT S. *The Effect of Air Pollution on Property Value in Chicago.* 1975. 111 p.
167. HANNEMANN, MANFRED. *The Diffusion of the Reformation in Southwestern Germany, 1518–1534.* 1975. 248 p.
168. SUBLETT, MICHAEL D. *Farmers on the Road. Interfarm Migration and the Farming of Noncontiguous Lands in Three Midwestern Townships. 1939–1969.* 1975. 228 pp.
169. STETZER, DONALD FOSTER. *Special Districts in Cook County: Toward a Geography of Local Government.* 1975. 189 pp.
170. EARLE, CARVILLE V. *The Evolution of a Tidewater Settlement System: All Hallow's Parish, Maryland, 1650–1783.* 1975. 249 pp.
171. SPODEK, HOWARD. *Urban-Rural Integration in Regional Development: A Case Study of Saurashtra, India—1800–1960.* 1976. 156 pp.

172. COHEN, YEHOSHUA S. and BERRY, BRIAN J. L. *Spatial Components of Manufacturing Change.* 1975. 272 pp.
173. HAYES, CHARLES R. *The Dispersed City: The Case of Piedmont, North Carolina.* 1976. 169 pp.
174. CARGO, DOUGLAS B. *Solid Wastes: Factors Influencing Generation Rates.* 1977. 112 pp.
175. GILLARD, QUENTIN. *Incomes and Accessibility. Metropolitan Labor Force Participation, Commuting, and Income Differentials in the United States, 1960–1970.* 1977. 140 pp.
176. MORGAN, DAVID J. *Patterns of Population Distribution: A Residential Preference Model and Its Dynamic.* 1978. 216 pp.
177. STOKES, HOUSTON H.; JONES, DONALD W. and NEUBURGER, HUGH M. *Unemployment and Adjustment in the Labor Market: A Comparison between the Regional and National Responses.* 1975. 135 pp.
179. HARRIS, CHAUNCY D. *Bibliography of Geography. Part I. Introduction to General Aids.* 1976. 288 pp.
180. CARR, CLAUDIA J. *Pastoralism in Crisis. The Dasanetch and their Ethiopian Lands.* 1977. 339 pp.
181. GOODWIN, GARY C. *Cherokees in Transition: A Study of Changing Culture and Environment Prior to 1775.* 1977. 221 pp.
182. KNIGHT, DAVID B. *A Capital for Canada: Conflict and Compromise in the Nineteenth Century.* 1977. 359 pp.
183. HAIGH, MARTIN J. *The Evolution of Slopes on Artificial Landforms: Blaenavon, Gwent.* 1978. 311 pp.
184. FINK, L. DEE. *Listening to the Learner. An Exploratory Study of Personal Meaning in College Geography Courses.* 1977. 200 pp.
185. HELGREN, DAVID M. *Rivers of Diamonds: An Alluvial History of the Lower Vaal Basin.* 1979. 399 pp.
186. BUTZER, KARL W., editor. *Dimensions of Human Geography: Essays on Some Familiar and Neglected Themes.* 1978. 201 pp.
187. MITSUHASHI, SETSUKO. *Japanese Commodity Flows.* 1978. 185 pp.
188. CARIS, SUSAN L. *Community Attitudes toward Pollution.* 1978. 226 pp.
189. REES, PHILIP M. *Residential Patterns in American Cities, 1960.* 1979. 424 pp.
190. KANNE, EDWARD A. *Fresh Food for Nicosia.* 1979. 116 pp.
191. WIXMAN, RONALD. *Language Aspects of Ethnic Patterns and Processes in the North Caucasus.* 1980. 224 pp.
192. KIRCHNER, JOHN A. *Sugar and Seasonal Labor Migration: The Case of Tucumán, Argentina.* 1980. 158 pp.
193. HARRIS, CHAUNCY D. and FELLMANN, JEROME D. *International List of Geographical Serials, Third Edition, 1980.* 1980. 457 p.
194. HARRIS, CHAUNCY D. *Annotated World List of Selected Current Geographical Serials, Fourth, Edition. 1980.* 1980. 165 p.
195. LEUNG, CHI-KEUNG. *China: Railway Patterns and National Goals.* 1980. 235 p.
196. LEUNG, CHI-KEUNG and NORTON S. GINSBURG, eds. *China: Urbanization and National Development.* 1980. 280 p.
197. DAICHES, SOL. *People in Distress: A Geographical Perspective on Psychological Well-being.* 1981, 199 p.
198. JOHNSON, JOSEPH T. *Location and Trade Theory: Industrial Location, Comparative Advantage, and the Geographic Pattern of Production in the United States.* 1981. 107 p.
199-200. STEVENSON, ARTHUR J. *The New York-Newark Air Freight System.* 1982. 440 p. (Double number, price: $16.00)
201. LICATE, JACK A. *Creation of a Mexican Landscape: Territorial Organization and Settlement in the Eastern Puebla Basin, 1520–1605.* 1981. 143 p.
202. RUDZITIS, GUNDARS. *Residential Location Determinants of the Older Population.* 1982. 117 p.
203. LIANG, ERNEST P. *China: Railways and Agricultural Development, 1875–1935.* 1982. 186 p.
204. DAHMANN, DONALD C. *Locals and Cosmopolitans: Patterns of Spatial Mobility during the Transition from Youth to Early Adulthood.* 1982. 146 p.
205. FOOTE, KENNETH E. *Color in Public Spaces: Toward a Communication-Based Theory of the Urban Built Environment.* 1983. 153 p.
206. HARRIS, CHAUNCY D. *Bibliography of Geography. Part II: Regional. Vol. 1. The United States of America.* 1984. 178 p.
207-208. WHEATLEY, PAUL. *Nāgara and Commandery: Origins of the Southeast Asian Urban Traditions.* 1983. 473 p.
(Double number, price: $16.00)
210. WESCOAT, JAMES L., JR. *Integrated Water Development: Water Use and Conservation Practice in Western Colorado.* 1984. 239 p.